Lecture Notes in Mathematics

2162

More information about this series at http://www.springer.com/series/304

Patrick Popescu-Pampu

What is the Genus?

Springer

Patrick Popescu-Pampu
UFR de Mathématiques
Université Lille 1
Villeneuve d'Ascq, France

Expanded translation by the author of the original French edition:
Patrick Popescu-Pampu, Qu'est-ce que le genre?, in: *Histoires de Mathématiques*, Actes
des Journées X-UPS 2011, Ed. Ecole Polytechnique (2012), ISBN 978-2-7302-1595-4,
pp. 55-198

ISSN 0075-8434   ISSN 1617-9692   (electronic)
Lecture Notes in Mathematics
ISBN 978-3-319-42311-1   ISBN 978-3-319-42312-8   (eBook)
DOI 10.1007/978-3-319-42312-8

Library of Congress Control Number: 2016950015

Mathematics Subject Classification (2010): 01A05, 14-03, 30-03, 55-03

Printed on acid-free paper

This Springer imprint is published by Springer Nature
The registered company is Springer International Publishing AG Switzerland

To Ghislaine, Fantin and Line

Preface to the English Translation

In France, some students follow special curricula during the first 2 years of their superior formation, in "classes préparatoires." There, an intensive training is organized for the entrance examinations to teaching institutions in science or engineering, the so-called "grandes écoles."

Every May, one of those "great schools," École Polytechnique, organizes a 2-day mathematical conference with lectures given by professional mathematicians and addressed to mathematics teachers of "classes préparatoires." Each year, the organizers choose a special theme.

In the beginning of 2011, Pascale Harinck, Alain Plagne, and Claude Sabbah invited me to give one of those lectures. The theme of that year was "Histoires de Mathématiques." This title has an ambiguity in French, as it may be understood both as "History of Mathematics" and "Stories about Mathematics." I chose to respect this ambiguity by speaking about the history of mathematics and at the same time by telling a story. The subject of this story was suggested to me by Claude Sabbah in his invitation message: "the notion of genus in algebraic geometry, arithmetic and the theory of singularities."

I accepted because I saw in the genus one of the most fascinating notions of mathematics, in its rich metamorphoses and in the wealth of phenomena it involves. It may be seen as the prototype of the concept of an invariant in geometry. Preparing the talk and writing the accompanying text for the proceedings to be published at the end of the same year appeared to me as an excellent opportunity to learn more about the development of this notion.

At that moment, I could not have imagined that navigating through the original writings of the discoverers would lead me to a book-length text! In it, I followed several of the evolutionary branches of the notion of genus, from its prehistory in problems of integration, through the cases of algebraic curves and their associated Riemann surfaces, then of algebraic surfaces, into higher dimensions. I had of course to omit many aspects of this incredibly versatile concept, but I hope that the reader who follows me will continue this exploration according to her or his own taste.

I am not a professional historian of mathematics, but I love to understand the development of mathematical ideas from this perspective. Such an understanding seems essential to me both for doing research and for communicating with other mathematicians or with students.

This book is a slightly expanded translation of the original French version [155]. I corrected a few errors; I reformulated several vague sentences; I added some explanations, figures, or references; and I reorganized the index. I also added two new chapters, one about Whitney's work on sphere bundles and another one on Harnack's formula relating the genus of a Riemann surface defined over the reals to the number of connected components of its real locus.

Acknowledgments I took great advantage from the teamwork leading to the book [52], especially the ensuing contact with writings of the nineteenth century. I want to thank all my co-authors. I am also keen to thank Clément Caubel, Youssef Hantout, Andreas Höring, Walter Neumann, Claude Sabbah, Michel Serfati, Olivier Serman, and Bernard Teissier for their help, their remarks, and their advice. I am particularly indebted to Maria Angelica Cueto for her very careful reading of the first version of my English translation and her advice for improving it. I am also very grateful to the language editor Barnaby Sheppard. Finally, I want to thank warmly Ute McCrory for having raised the idea to publish this text as a book in the History of Math subseries of Springer Lecture Notes in Mathematics.

Villeneuve d'Ascq, France   Patrick Popescu-Pampu

Contents

Introduction

Nowadays, one of the fastest ways to introduce the mathematical notion of *genus* is probably to say that it is the number of holes of a surface. If one is speaking to a person with enough mathematical education, one has to add that this surface should be compact, connected, orientable, and without boundary. For instance (see Fig. 1), a sphere is of genus 0, a torus is of genus 1, and the surface of a pretzel is of genus 3.

This definition has the advantage of being intuitive: one may explain it through examples even to children. Moreover, with a little training, one can rapidly manage to find the genus of a given surface provided that it is not too twisted or knotted, as in Fig. 2,[1] which shows only surfaces of genus 0, or as in Fig. 3, which shows a surface of genus 5.

The examples of this type enable us to understand that the concept of "hole" is not always meaningful. Is there some other concept, perhaps less intuitive, which could be applied to any surface and which would give the number of holes whenever possible, for instance, for the surfaces of Fig. 1?

Over the last two centuries, many mathematicians have tried to define a concept of "genus" which is applicable to all surfaces, possibly located in spaces of higher dimension, and even to "abstract" surfaces, which are not given inside any ambient space different from themselves.

Let us see how one may arrive at such a definition, which no longer refers to an ambient space. Start from intuitive examples, where the holes are immediately recognizable. Then draw contours which surround those holes on the surface. Since the holes are separated, one may choose those contours to be pairwise disjoint. One comes up with a collection of circles drawn on the surface, exactly as many as the number of holes, as illustrated in Fig. 4.

We have found an idea: draw pairwise disjoint circles on any surface, then count them, and say that their number is the *genus* of the surface. In order to transform this construction into a well-defined concept, one has to explain first under which

---

[1]Photograph of a runic stone taken in the city of Sigtuna (Sweden) in 1914 by Erik Brate and available at http://commons.wikimedia.org/wiki/File:U_460,_Skr.

Fig. 1 A sphere, a torus, and two pretzels

constraints one must choose the circles and secondly that all such choices give the same number of circles.

One could either choose a single circle, or one could keep drawing circles, each time slightly different from the circles already drawn. In order to understand how to forbid such choices of circles, which would not allow one to arrive at a uniquely defined number, consider again one of the initial examples, in which a circle surrounds each hole. Then cut the surface along these circles. One sees that the new surface remains connected. But, as indicated by as many examples as one desires, adding an extra circle and performing one more cut would disconnect the surface.

One arrives at the following definition:

The genus of a (compact, connected, orientable) surface without boundary is the maximal number of pairwise disjoint circles one can draw on the surface, with connected complement.

It is then a theorem that all the sets of circles which satisfy those constraints have the same number of elements. This definition applies to all abstract surfaces, as it uses only constructions performed inside the surface, without any reference to an ambient space.

Of course, in order to get a definition which is perfectly satisfying not only from the intuitive viewpoint but also logically, one has to define precisely the notions of surface, of circle drawn on it, of cutting along such a circle, and of connectedness. Topology was developed in particular in order to give a meaning to all these concepts. If one then carefully proves the previously stated theorem of

Fig. 2 A runic stone

invariance of the number of circles, one gets indeed a concept of "genus" which is rigorously constructed from a logical viewpoint.

But this does not explain the reason why this concept emerged, nor why it is relevant. In fact, its importance comes from its many avatars, each one of them suggesting other generalizations in higher dimensions, and from the fact that all those generalizations are the basic characteristics used to classify geometric beings in analogy with the classification of living beings.

We will examine here various expressions of this concept during our stroll through time. Exhaustiveness is not an aim of this stroll; it is simply an invitation to listen to the mathematicians of the past. I chose to present many citations, in order to let the actors speak about their motivations and several spectators about their interpretations. In this way, the variety of styles gets emphasized, as well as the evolution of the language, of the questions, and of the viewpoints.

This stroll has three parts: in the first one, we deal with algebraic curves and their topological manifestation once we look at their complex points, forming

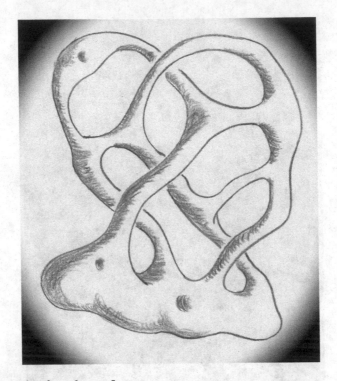

Fig. 3 A knotted surface of genus 5

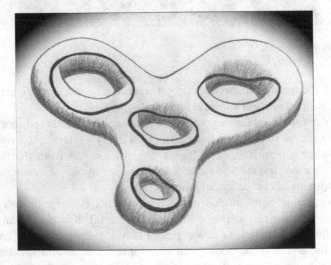

Fig. 4 Contours surrounding the holes

Riemann surfaces. The second part examines the diverse notions of genus which were introduced for algebraic surfaces. Finally, in the last part, we examine generalizations to arbitrary finite dimensions. But, before starting into this journey, we shall see how Aristotle explained the meaning of the term "$\gamma \acute{\epsilon} \nu o \varsigma$."

Chapter 1
The γένος According to Aristotle

The term *genus* reached us from the ancient Greek γένος, via Latin. It is a term already used for classifications during the times of Aristotle, as illustrated by the following extract from his *"Metaphysics"* [5, Book 5, Chap. 1024]:

> The term "genus" (or "race") is used: (a) When there is a continuous generation of things of the same type; e.g., "as long as the human race exists" means "as long as the generation of human beings is continuous." (b) Of anything from which things derive their being as the prime mover of them into being. Thus some are called Hellenes by race, and others Ionians, because some have Hellen and others Ion as their first ancestor. [...] (c) In the sense that the plane is the "genus" of plane figures, and the solid of solids [...] (d) In the sense that in formulae the first component, which is stated as part of the essence, is the genus, and the qualities are said to be its differentiae. [...]
>
> Things are called "generically different" whose immediate substrates are different and cannot be resolved one into the other or both into the same thing. E.g., form and matter are generically different, and all things which belong to different categories of being; for some of the things of which being is predicated denote the essence, others a quality, and others the various other things which have already been distinguished. For these also cannot be resolved either into each other or into any one thing.

© Springer International Publishing Switzerland 2016
P. Popescu-Pampu, *What is the Genus?*, Lecture Notes in Mathematics 2162,
DOI 10.1007/978-3-319-42312-8_1

Part I
Algebraic Curves

Chapter 2
Descartes and the New World of Curves

Let us make a huge temporal leap, in order to reach the *"Géométrie"* [56] of Descartes, published in 1637, illustrating his *"Discourse on the Method of Rightly Conducting One's Reason and of Seeking Truth in the Sciences"*.

In that text, Descartes unveiled a new world of *curves*. He observed that the conical sections of the ancients, once related to a pair of intersecting lines which are moreover endowed with a unit of measurement (which is called, in memory of him, *"a system of Cartesian coordinates"*), may all be described by a polynomial equation of degree two. He asserted then that one should also study the curves defined by equations of higher degree. However, the ancients did not undertake such a study, except in some particular cases (see Brieskorn and Knörrer's book [25, Sect. I.1]). Descartes explained this fact in the following way [56, Book Two, page 44][1]:

> Probably the real explanation of the refusal of ancient geometers to accept curves more complex than the conic sections lies in the fact that the first curves to which their attention was attracted happened to be the spiral, the quadratix, and similar curves, which really belong only to mechanics, and are not among those curves that I think should be included here, since they must be conceived of as described by two separate movements whose relation does not admit of exact determination. Yet they afterwards examined the conchoid, the cissoid, and a few others which should be accepted; but not knowing much about their properties, they took no more account of these than of the others. Again, it may have been that, knowing as they did only a little about the conic sections, and being still ignorant of many of the possibilities of the ruler and compasses, they dared not yet attack a matter of still greater difficulty. I hope that hereafter those who are clever enough of the geometric methods herein suggested will find no great difficulty in applying them to plane or solid problems. I therefore think it proper to suggest to such a more extended line of investigation which will furnish abundant opportunities for practice.

We discover here a Descartes eager to establish the frontiers of his new world of curves: some of them, which he called "mechanical" (for instance the Spiral), do not belong to it. We will see in the next chapter that, instead, Newton included some of

---

[1] The translation into English is taken from [56].

© Springer International Publishing Switzerland 2016
P. Popescu-Pampu, *What is the Genus?*, Lecture Notes in Mathematics 2162,
DOI 10.1007/978-3-319-42312-8_2

the mechanical curves among the "geometrical" ones, but inside a special category. Namely, that of the curves of *infinite degree*.

In fact, Descartes rarely used the term *"degree"*, and if he did, then only in a metaphorical way, in order to speak about a gradation in the complexity of curves. He kept interpreting the unknowns in the Ancients' way, as lengths of segments. Consequently, a monomial which is for us of *degree d*, for him corresponded to the volume of a parallelepiped of *dimension d*. Therefore, he arranged polynomial equations with two variables according to their *dimensions*. For instance, here is what he wrote in [56, Book Two, page 48]:

> I could give here several other ways of tracing and conceiving a series of curved lines, each curve more complex than any preceding one, but I think the best way to group together all such curves and then classify them in order by certain genera,[2] is by recognizing the fact that all points of those curves which we may call "geometric", that is, those which admit of precise and exact measurement, must bear a definite relation to all points of a straight line, and that this relation must be expressed by means of a single equation. If this equation contains no term of higher degree than the rectangle of two unknown quantities, or the square of one, the curve belongs to the first and simplest genus,[3] which contains only the circle, the parabola, the hyperbola, and the ellipse; but when the equation contains one or more terms of the third or fourth degree in one or both of the two unknown quantities (for it requires two unknown quantities to express the relation between two points) the curve belongs to the second one; and if the equation contains a term of the fifth or sixth degree in either or both of the unknown quantities the curve belongs to the third class, and so on indefinitely.

We may notice that Descartes asserts that, in this way, he arranged the curves by *"genera"*. This is not yet the present-day meaning. What is unusual is to see him bring together in the same n-th "genus" the curves defined by equations of degrees $2n - 1$ and $2n$. The reason is rather mysterious [56, page 56]:

> [...] there is a general rule for reducing to a cubic any equation of the fourth degree, and to an equation of the fifth degree any equation of the sixth degree, so that the latter in each case need not be considered any more complex than the former.

Here is what Michel Serfati, a specialist in Descartes' work, wrote me about this issue: *"Descartes indicates first that the 4-th degree may be reduced to the 3-rd. This conclusion will be established algebraically in the book III, by classically bringing a 4-th degree equation to a 3-rd degree resolvent. The method is interesting and specific to him, different from Ferrari's one from the* Ars Magna *of 1545, a text which we are sure that Descartes knew. [...] Starting from this situation, Descartes believes that he may state, without proof and by a false extension, that the curves of the third genus (5-th and 6-th degrees) may all be reduced to the 5-th one, which would therefore represent them all"*.

---

[2] Smith and Latham translated simply as "classify them in order". Nevertheless, the French original says "les distinguer par ordre en certains genres".
[3] Smith and Latham translated the original "premier et plus simple genre" as "first and simplest class".

Chapter 3
Newton and the Classification of Curves

During his youth, Newton had carefully studied the geometric calculus of Descartes, which served him as a source of inspiration for the development of the "calculus of fluxions", his version of the differential calculus. This partially explains why he undertook to classify the curves of degree three according to various species, in analogy with the classification of those of degree two, the *conic sections*, into *ellipses, parabolas, hyperbolas* or *pairs of lines*. The following is the first paragraph of the chapter containing this classification from his work [140], published in 1711:

> Geometrical lines are best divided into orders, according to the dimensions of the equation expressing the relation between absciss and ordinate, or, which is the same thing, according to the number of points in which they can be cut by a straight line. So that a line of the first order will be a straight line; those of the second or quadratic order will be conic sections and the circle; and those of the third or cubic order will be the cubic Parabola, the Neilian Parabola, the Cissoïd of the ancients, and others we are about to describe. A curve of the first genus (since straight lines are not to be reckoned among curves) is the same as a line of the second order, and a curve of the second genus is the same as a line of the third order. And a line of the infinitesimal order is one which a straight line may cut in an infinite number of points, such as the spiral, cycloïd, quadratrix, and every line generated by the infinitely continued rotations of a radius.

We see that Newton wrote about "*genera*" in relation to "curves" but about "*orders*" in relation to "lines". His notion of *genus* differs from Descartes' one, as a polynomial of degree n defines a curve of *order* n and a line of *genus* $n - 1$. It seems strange to see him using two distinct but equivalent terms to speak about the same objects. It is probable that he wanted to use the two standard terms of his times, and that common language was reluctant to say that *a straight line was curved*.

Notice also the geometric interpretation of the degree of a curve, as *the number of intersection points with a straight line*. This is of course to be interpreted cautiously, as was understood later. Namely, if one wants to get an equality for all curves, one has to look not only at the real intersection points, but also at the complex ones, to consider only certain lines (neither the asymptotes, nor those directed towards

© Springer International Publishing Switzerland 2016   7
P. Popescu-Pampu, *What is the Genus?*, Lecture Notes in Mathematics 2162,
DOI 10.1007/978-3-319-42312-8_3

asymptotic directions), and to count with adequate multiplicities the intersection points.

All these aspects were to be progressively clarified thanks, on one hand, to the proof of the "*fundamental theorem of algebra*" and the consideration of the complex points of the plane, and on the other hand to the addition of the "*straight line at infinity*", which enabled the intersection points to be at infinity. This amounts to work in the *complex projective plane*, which was the privileged environment for the geometric study of algebraic curves in the nineteenth century (see, for instance, Stillwell's historical book [172] as well as the historical information contained in Brieskorn and Knörrer's book [25]).

Chapter 4
When Integrals Hide Curves

In the two previous chapters we dealt with curves and the polynomials defining them or the mechanisms generating them. Those curves often represented incarnations of problems involving polynomial equations, with one or more variables.

In the last quarter of the seventeenth century, Newton and Leibniz developed in distinct ways the foundations of differential and integral calculus, which triggered a famous priority dispute. At the end of that century a new type of problem had appeared, that of explicit integration of the differentials $f(x)dx$. That is, the problem of computing primitives $\int f(x)dx$, where $f(x)$ denotes a given function.[1] In fact, *this problem is also related to the study of curves*!

In order to explain this, let us start from the following exercise contained in the lessons of integral calculus given by Johann Bernoulli to the marquis of l'Hôpital [19, page 393], and picked up again with modern notations by André Weil [190, page 400]:

> Everything amounts therefore to rendering irrational expressions rational...a task in which the Diophantine questions are of a great help...For instance, if one wants to integrate
>
> $$\frac{a^2 dx}{x\sqrt{ax - x^2}};$$
>
> one will perform the change of variables $ax - x^2 = a^2x^2t^{-2}\ldots$

In modern language, the problem posed by Bernoulli is that of the computation of the primitives of the function $\dfrac{a^2}{x\sqrt{ax - x^2}}$. The change of variable proposed by

---

[1]Incidentally, the term *"function"* was introduced in this context by Leibniz.

© Springer International Publishing Switzerland 2016
P. Popescu-Pampu, *What is the Genus?*, Lecture Notes in Mathematics 2162,
DOI 10.1007/978-3-319-42312-8_4

him allows one to express dx in terms of dt, which yields:

$$\int \frac{a^2 dx}{x\sqrt{ax-x^2}} = -2\left(\frac{a^2}{t}+t\right) + const$$

$$= -2a\left(\sqrt{\frac{x}{a-x}} + \sqrt{\frac{a-x}{x}}\right) + const.$$

In particular, the integral *is again an algebraic function of the variable x* (that is, a function $y(x)$ which satisfies a polynomial equation $P(x, y) = 0$), as was the case for its integrand. We will come back in more detail to the notion of *algebraic function* in Chap. 12.

Which curve is hidden behind the previous computation? In order to find it, notice that the function to be integrated is not a rational function, as it contains a square root. Let us introduce a new variable equal to this square root: $y = \sqrt{ax-x^2}$. This equation becomes $x^2 - ax + y^2 = 0$, which indeed defines a curve in the plane of cartesian coordinates (x, y). Using the usual measure of length, we have the equation of a *circle* passing through the origin. The previous process of integration is based on the fact that a circle *may be rationally parametrized*, that is, parametrized using rational functions. Concretely, we have:

$$\begin{cases} x = \dfrac{at^2}{a^2 + t^2} \\ y = \dfrac{a^2 t}{a^2 + t^2} \end{cases}.$$

This process of integration may be applied each time one starts from a differential expression of the form $F(x, \sqrt{q(x)})dx$, where $q(x)$ is a polynomial of degree two in x and $F(u, v)$ is a rational function in two variables. Indeed, the associated curve is the one defined by the equation $y^2 - q(x) = 0$, which is again a conic section. But conic sections may always be rationally parametrized, by projecting them stereographically onto a line from one of their points. This allows us to transform the previous integral by a change of variable into the integral of a rational function.

Recall now the theorem of decomposition of rational functions into *simple rational functions*, that is, as sums of monomials and of fractions of the form $\frac{b}{(x-a)^n}$, with $a, b \in \mathbb{C}$ and $n \in \mathbb{N}^*$. It was developed precisely in this context of computation of primitives of rational functions, and the belief in its generality was an important stimulus to prove the fundamental theorem of algebra (see Houzel [105, Chap. III]). One gets:

Theorem 4.1 *If F is a rational function in two variables, then the primitives $\int F(x, \sqrt{q(x)})dx$ are sums of algebraic functions of the variable x and of logarithms of such functions.*

Chapter 5
Jakob Bernoulli and the Construction of Curves

In the following extract of [17],[1] Jakob Bernoulli, the brother of Johann, about whom we wrote in Chap. 4, analyzed various methods of construction of "mechanical" or "transcendental" curves, that is, curves which are not "algebraic" (defined by a polynomial equation). Those methods created a common framework for Descartes' algebraic curves and for the curves furnished by the differential and integral calculus:

> There are three main procedures to construct mechanical or transcendental curves. The first one consists in the quadrature of curvilinear areas, but it is poorly adapted to practice. It is better to rectify algebraic curves; because in practice it is easier and more precise to rectify curves, using a thread or a small chain wrapping the curve, than to square a surface. I equally appreciate the constructions which proceed without any rectification or quadrature, by the simple description of a mechanical curve for which it is possible to determine geometrically, even if not all of them, at least an infinite number of points which are arbitrarily close of each other; one finds among them the logarithmic curve and perhaps also other curves of the same kind. But the best method, as far as it is applicable, is the one using a curve which nature produces itself without any trick, by a rapid and almost instantaneous movement, at the first glance of the geometer. Because all the methods cited before need curves whose construction—done by a continuous movement or by the invention of several points—is usually too slow and too laborious. That is why I believe that the constructions in the problems which suppose the quadrature of a hyperbola or the description of a logarithmic curve are *ceteris paribus* less favourable than those which are done using a catenary: because a chain will take by itself this shape before one is able even to start the construction of the other ones.

Let us rephrase this in modern terms. When one starts from a known function $f(x)$, one of its primitives $\int f(x)dx$ is a new function. In geometric terms, if the graph of $f(x)$ is a known curve, then the graph of $\int f(x)dx$ is a new curve. This is Bernoulli's first "procedure" by "quadratures". His second procedure starts from a known curve seen as the graph of a function $f(x)$, and takes the graph of the

---

[1]I found this extract in the paper [171, Sect. 2] of Smadja, which is my main source of information for the content of this chapter.

P. Popescu-Pampu, *What is the Genus?*, Lecture Notes in Mathematics 2162,

DOI 10.1007/978-3-319-42312-8_5

function which associates to x the length of the arc going from the point with abscissa x to a point of fixed abscissa. In fact, computing the length of an arc was called "rectifying" it (concretely speaking, as explained by Jakob Bernoulli, one may straighten "a small chain wrapping the curve" simply by stretching it).

If one cannot calculate an integral as in Theorem 4.1, one may try to view it as the integral associated with the rectification of a known curve, for instance, an algebraic curve.

It was while dealing with this kind of question that the two brothers encountered, in seemingly independent ways, the same curve in 1694 (their publications on this subject being [17] and [18]). They both tried to solve *the problem of the paracentric isochrone*, which had been formulated by Leibniz: "*Find the curve on which the fall of a heavy body makes it move away or brings it closer to a given point in a uniform way.*" They reduced the problem to the computation of the integral:

$$\int \frac{dx}{\sqrt{1 - x^4}}, \tag{5.1}$$

and they realized that this integral also arises in the problem of rectifying the curve with equation:

$$(x^2 + y^2)^2 = x^2 - y^2. \tag{5.2}$$

This curve, the *lemniscate*, is represented in Fig. 5.1. It may also be defined geometrically as *the locus of points in a plane whose product of distances to two fixed points is the same as in the case of the midpoint of the segment joining the two given points.*

The name "*lemniscate*" was coined by Jakob Bernoulli in [17], as the curve has "*the shape of a reversed figure eight* ∞, *of a knotted loop, of a* λημνίσκοσ, *or, in English, of a* ribbon knot".

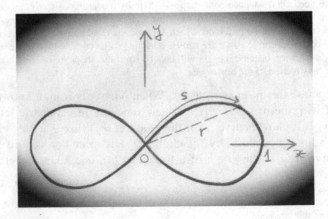

Fig. 5.1 Measuring arc-length on a lemniscate

In order to see that the integral (5.1) encodes the problem of rectification of the lemniscate, one may use polar coordinates. More precisely, one expresses x^2 and y^2 in terms of the distance r to the origin using Eq. (5.2). Then, by differentiation and a small computation, one arrives at the following expression for the length s of the arc which goes from the origin to a point situated on the lemniscate in the positive quadrant (see Fig. 5.1):

$$s = \int \sqrt{dx^2 + dy^2} = \int \frac{dr}{\sqrt{1 - r^4}}. \tag{5.3}$$

The expression under the square root in (5.3) having degree four, the associated curve, with equation $y^2 = 1 - x^4$, is no longer a conic. Therefore, the method of computation explained in the previous chapter fails.

Can one solve the problem in a different way? The brothers Bernoulli could not answer this question. Only after the works of Fagnano one could begin to say something about this "lemniscatic integral", even without being able to compute it explicitly.

Chapter 6
Fagnano and the Lemniscate

Fagnano published his study of the lemniscate in his 1718 papers [56] and [57]. He was so proud of it that, in the 1750 edition of his collected mathematical works, he chose a lemniscate for the decoration of the front page, surmounted by the motto *"Deo Veritatis Gloria"* (see Fig. 6.1).

Here is how Fagnano described his research on the lemniscate in the paper [58]:

> Two great geometers, Jakob and Johann Bernoulli, made famous the lemniscate by using its arcs in order to construct the paracentric isochrone, as may be seen in the Acts of Lipse of 1694. It is clear that, by measuring the lemniscate using a simpler curve, one obtains a better construction not only of the paracentric isochrone, but also of an infinite number of other curves whose constructions may depend on the lemniscate; that is why I pride myself that the measures of this curve which I discovered and which I will present in two short memoirs will please those who understand something about this subject.

Here we will examine only one of Fagnano's formulae, the one giving the *doubling of the arc of the lemniscate* (see [171, Sect. 4]):

Theorem 6.1 *If* $\dfrac{u\sqrt{2}}{\sqrt{1-u^4}} = \dfrac{1}{z}\sqrt{1 - \sqrt{1-z^4}}$, *then:*

$$\frac{dz}{\sqrt{1-z^4}} = \frac{2du}{\sqrt{1-u^4}}.$$

This may be rewritten in the following way, by first expressing z in terms of u:

$$z = \frac{2u\sqrt{1-u^4}}{1+u^4} \implies 2\int_0^u \frac{dt}{\sqrt{1-t^4}} = \int_0^z \frac{dt}{\sqrt{1-t^4}}. \tag{6.1}$$

This implication shows that, if one starts from a point of the lemniscate situated at distance u from the origin then, in order to double the corresponding arc $s(u)$, one may take a point situated at distance z from the origin, u and z being related by an explicit algebraic equation.

© Springer International Publishing Switzerland 2016
P. Popescu-Pampu, *What is the Genus?*, Lecture Notes in Mathematics 2162,
DOI 10.1007/978-3-319-42312-8_6

PRODUZIONI
MATEMATICHE
DEL CONTE GIULIO CARLO
DI FAGNANO,
MARCHESE DE' TOSCHI,
E DI SANT' ONORIO
NOBILE ROMANO, E PATRIZIO SENOGAGLIESE

ALLA SANTITA' DI N.S.

BENEDETTO XIV.
PONTEFICE MASSIMO.
TOMO PRIMO.

IN PESARO
L' ANNO DEL GIUBBILEO M. DCC. L.
NELLA STAMPERIA GAVELLIANA
CON LICENZA DE' SUPERIORI.

Fig. 6.1 Do you see the lemniscate?

One may find a comparison of the previous discussion with the formulae for doubling the arcs of a circle in Stillwell's book [172, 12.4] and a detailed description of Fagnano's works on the lemniscate in Smadja's paper [171, Sect. 2].

Chapter 7
Euler and the Addition of Lemniscatic Integrals

In 1751, Fagnano applied to become a member of the Berlin academy of sciences. It was on that occasion that Euler studied his works, which gave him new ideas. He arrived in 1753 at the following generalization of formula (6.1):

Theorem 7.1 *If* $c = \dfrac{a\sqrt{1 - b^4} + b\sqrt{1 - a^4}}{1 + a^2b^2}$, *then:*

$$\int_0^a \frac{dt}{\sqrt{1 - t^4}} + \int_0^b \frac{dt}{\sqrt{1 - t^4}} = \int_0^c \frac{dt}{\sqrt{1 - t^4}}.$$

Note that one recovers (6.1) by setting $a = b$ in Euler's formula.

In [73], before describing his research triggered by the study of Fagnano's works on the lemniscate, including the "*addition formula*" above, Euler presented his vision on the usefulness of mathematical "speculations":

If one examines their usefulness, mathematical speculations seem to be reducible to two great classes; to the first class belong those who bring a remarkable benefit both to the common life and to the other arts, and the price grows usually with this benefit. But another class collects those speculations which, without being related to any remarkable benefit, are nevertheless able to develop the ends of Analysis and to sharpen the strength of our spirit. Indeed, as we are led to abandon a great many research from which we may expect a lot of profit, only because of a want of analysis, it seems that we do not have to estimate at a lower price a work which promises non-negligible advances in Analysis. With this goal in mind, those observations are particularly valuable, which were done almost by chance and were discovered a posteriori, for which one sees neither any a priori reason nor any direct way to reach them. But the truth being already known, it will be possible to search more easily among those observations for some methods which lead directly to that truth. It is clearly beyond any doubt that the search for new methods contributes a lot to the promotion of the goals of Analysis.

In the recently published book of the count Fagnano I found several observations of this sort which were done without a clear method and whose explanation seems rather hidden. [...]

© Springer International Publishing Switzerland 2016
P. Popescu-Pampu, *What is the Genus?*, Lecture Notes in Mathematics 2162,
DOI 10.1007/978-3-319-42312-8_7

As the reason for those properties seems deeply hidden, I believe that it would not be inopportune to examine them in a more direct way, and to share with the public the properties which I was lucky to discover concerning those curves.

We see here Euler's eagerness to rise from the particular to the general, to extract from an experimental material (here the writings of Fagnano) a method which may be applied to the largest possible contexts. Thus he succeeded to generalize the addition formula of Theorem 7.1 to the case where $1 - x^4$ is replaced by an arbitrary polynomial of the fourth degree (see [105, VIII.3]).

Having reached this level of generality, in which the lemniscate became indiscernible from the plane curves of degrees 3 or 4, we will leave the lemniscate, coming back to it only in order to illustrate general theorems. The reader eager to learn much more about this special curve may consult the historical papers [11, 163, 171] of Ayoub, Schappacher and Smadja, respectively, as well as the last chapter of Cox's textbook [48] on Galois theory.

Chapter 8
Legendre and Elliptic Functions

Starting in the late 1700s, Legendre spent several decades developing a general theory of the integrals which he called *elliptic functions*, and which satisfy an addition formula analogous to that established by Euler in Theorem 7.1. Here is what he wrote about his motivations in the foreword of his 1825 book [128]:

> Euler had predicted that with the aid of adapted notations, the computation of arc-lengths of ellipses and of other analogous transcendents[1] could become as standard a method as that of arcs of circles and of logarithms[2]; but, with the exception of *Landen*, who, by the discovery of his theorem could have paved new roads, nobody tried to fulfil Euler's prediction, and one may say that the Author of the present Treatise remained alone in dealing with it, since the year 1786 when he published his first research on the arcs of ellipses, till the present time.

Legendre returned to this theme on the first page of his Introduction:

> If it were possible to arrange methodically the various transcendents which were known and used till now only under the name of *quadratures*; if by studying their properties one could reduce them to the simplest possible expressions compatible with their degrees of generality, and to compute easily their approximate values when they become completely determined; then the corresponding transcendents, designated by individual letters and submitted to an adequate algorithm, could be used in analysis similarly to the arcs of circles and the logarithms; the applications of the integral calculus would not be stopped any more, as it was the case till now, by this sort of barrier which nobody tries to overcome, when the problem is brought to quadratures and the solutions, barely begun by this reduction, would receive all the developments allowed by the nature of the question.

---

[1] That is, non-algebraic functions.

[2] Here are the exact words of Euler (Novi Com. Petrop., tom. X, pag. 4): "*Imprimis autem hic idoneus signandi modus desiderari videtur, cujus ope arcus elliptici æque commode in calculo exprimi queant ac jam logarithmi et arcus circulares, ad insigne analyseos incrementum, in calculum sunt introducti. Talia signa novam quamdam calculi speciem supeditabunt.*"

© Springer International Publishing Switzerland 2016

P. Popescu-Pampu, *What is the Genus?*, Lecture Notes in Mathematics 2162,

DOI 10.1007/978-3-319-42312-8_8

What would be impossible to achieve in a frame so broad as the one just described, can at least be realized for the transcendents which are closest to circular and logarithmic functions, as are the arcs of ellipses and hyperbolas, and in general the transcendents which we called *elliptic functions*.

Let us see how Legendre defined those elliptic functions, and how he explained his choice of name in [128, Chap. V]:

Because the transcendents we considered may be always reduced to the two forms

$$\int (A + B \sin^2 \phi) \frac{d\phi}{\Delta}, \ \int \frac{d\phi}{(1 + n \sin^2 \phi)\Delta}$$

[where $\Delta := 1 - c^2 \sin^2 \phi$, with $-1 \leq c \leq 1$ and A, B, n are parameters], they are clearly encompassed by the general formula

$$H = \int \frac{A + B \sin^2 \phi}{1 + n \sin^2 \phi} \cdot \frac{d\phi}{\Delta}.$$

From now on we will call the integrals encompassed by this formula *elliptic functions* or *transcendents*. [...]

It does not seem plausible that the function H, taken in its full generality, could be reduced to arcs of ellipses; this takes place only when $n = 0$, or when some substitution allows to remove the denominator $1 + n \sin^2 \phi$, which we do not think will be possible in general. This shows that the denomination of elliptic function is inappropriate in some respect; however we choose to adopt it, due to the great analogy which will be found between the properties of this function and those of the arcs of ellipses.

Chapter 9
Abel and the New Transcendental Functions

At the time when Legendre published his treatise [128], Abel, a young Norwegian mathematician, was beginning to publish a huge generalization of the addition theorems for elliptic functions. This generalization dealt with *all* the integrals of the form:

$$\int y \, dx, \tag{9.1}$$

y being an *arbitrary* algebraic function of the variable x. Those integrals were later called "*abelian transcendents*" by Jacobi in [107]. In the sequel, we use the simpler name of *abelian integrals*.

For Abel, it was important to study those integrals, because they enriched the catalogue of functions which may be useful in Mathematics. Indeed, this may be seen already in the introduction to his 1826 article [1]:

> Until now, only a very small number of transcendental functions have been considered by the geometers. Almost all the theory of transcendental functions reduces to that of logarithmic, exponential and circular ones, functions which fundamentally constitute a single species. It is only in recent times that other kinds of functions have been considered. Among them, the elliptic transcendents, of which Legendre developed so many remarkable properties, are at the top of the list. In the memoir which he is honored to present to the Academy, the present author considered a very large class of functions, namely, those whose derivatives may be expressed using algebraic equations, all of whose coefficients are rational functions of the same variable. He found for those functions properties which are similar to those of logarithmic or elliptic functions.
>
> A function whose derivative is rational has the known property that one may express the sum of an arbitrary number of such functions using an algebraic and a logarithmic function [...]. Similarly, an arbitrary elliptic function, that is, a function whose derivative does not contain any other irrationality than a radical of degree two, under which the variable does not exceed the fourth degree,[1] will also have the property that one may express

---

[1] That is, the square root of a polynomial of degree at most four.

© Springer International Publishing Switzerland 2016
P. Popescu-Pampu, *What is the Genus?*, Lecture Notes in Mathematics 2162,
DOI 10.1007/978-3-319-42312-8_9

an arbitrary sum of such functions by an algebraic and a logarithmic function, provided that one establishes between the variables of those functions a certain algebraic relation. This analogy between the properties of those functions led the author to search whether it is possible to find analogous properties of more general functions, and he arrived at the following theorem:

"If one has several functions with derivatives which are roots of the *same algebraic equation*, all of whose coefficients are *rational* functions of the same variable, one may always express the sum of an arbitrary number of such functions by an *algebraic* and a *logarithmic* function, provided that one establishes a certain number of *algebraic* relations among the variables of the corresponding functions."

The number of such relations does not depend in any way on the number of functions, but only on the nature of the particular functions considered. For instance, for an elliptic function this number is 1; for a function whose derivative does not contain any other irrationality than a radical of degree two, under which the variable does not exceed the fifth or the sixth degree, the number of needed relations is 2, and so on.

This theorem is rather mysterious, and it was already so to Abel's contemporaries (see the historical information given by Kleiman in [115]). I will give a modern interpretation of it in Chap. 17.

In any case, *it is here that the notion of genus appears for the first time in the sense that interests us*! As a number which pops up while counting the relations which have to be imposed in order to arrive at a certain kind of identity concerning abelian integrals.

For this reason, one may consider that the history of the understanding of what will be called *the genus* (of the curve associated with the algebraic function $y(x)$ which appears in the abelian integral) begins with this paper of Abel.

Chapter 10
A Proof by Abel

Confronted with his contemporaries' difficulties in understanding his very general ideas, Abel decided to present some of them in separate articles. For instance in 1829, in the paper [3], he formulated the following version of the theorem cited in extenso in the previous chapter. This version makes no reference to the "number of algebraic relations" (and therefore to the genus):

Theorem *Let y be a function of x which satisfies some irreducible equation of the form*

$$0 = p_0 + p_1 y + p_2 y^2 + \cdots + p_{n-1} y^{n-1} + y^n, \tag{10.1}$$

where $p_0, p_1, p_2, \ldots, p_{n-1}$ are entire functions [that is, polynomials] of the variable x. Likewise, let

$$0 = q_0 + q_1 y + q_2 y^2 + \cdots + q_{n-1} y^{n-1}, \tag{10.2}$$

be a similar equation, where $q_0, q_1, q_2, \ldots, q_{n-1}$ are also entire functions of x. Suppose that the coefficients of the various powers of x in these functions are variable. We denote those coefficients by $a, a', a'' \ldots$ According to the two Eqs. (10.1) and (10.2), x will be a function of $a, a', a'' \ldots$ and we can determine its values eliminating the quantity y. Denote by

$$\rho = 0 \tag{10.3}$$

the result of the elimination. Therefore ρ contains only the variables $x, a, a', a'', \ldots$ Let μ be the degree of this equation relative to x and denote by

$$x_1, x_2, x_3, \ldots, x_\mu \tag{10.4}$$

its μ roots, which will be functions of $a, a', a'', \ldots$ This being established, if one sets

$$\psi x = \int f(x, y) dx \tag{10.5}$$

where $f(x, y)$ denotes an arbitrary rational function of x and y, I say that the transcendental function ψx will enjoy the general property expressed by the following equation:

$$\psi x_1 + \psi x_2 + \cdots + \psi x_\mu = u + k_1 \log v_1 + k_2 \log v_2 + \cdots + k_n \log v_n, \tag{10.6}$$

© Springer International Publishing Switzerland 2016
P. Popescu-Pampu, *What is the Genus?*, Lecture Notes in Mathematics 2162,
DOI 10.1007/978-3-319-42312-8_10

where $u, v_1, v_2, \ldots, v_n$ are rational functions of $a, a', a'', \ldots$, and $k_1, k_2, \ldots, k_n$ are constants.

Let us explain the reason why there is no reference in the previous statement to the number of relations between the variables. Both this statement and the one quoted in the previous chapter refer to *"functions whose derivatives may be roots of the same algebraic equation"*. Here, those functions are $\psi x_1, \ldots, \psi x_\mu$. In both cases, one considers their sum. But if in the first statement the variables were $x_1, \ldots, x_\mu$, in the second one the auxiliary variables $a, a', a'', \ldots$ determine $x_1, \ldots, x_\mu$ through the relations (10.1) and (10.2).

Abel's proof of the previous theorem is based on a clever use of the fact that any symmetric polynomial can be expressed as a polynomial in the elementary symmetric functions:

Proof In order to establish this theorem, it is enough to express the differential of the first member of the Eq. (10.6) as a function of $a, a', a'', \ldots$; because it will reduce in this way to a rational differential, as we will see. First, the two Eqs. (10.1) and (10.2) will give y as a rational function of $x, a, a', a'', \ldots$ Similarly, the Eq. (10.3) $\rho = 0$ will give an expression for dx of the form

$$dx = \alpha.da + \alpha'.da' + \alpha''.da'' + \cdots,$$

where $\alpha, \alpha', \alpha'', \ldots$ are rational functions of $x, a, a', a'', \ldots$ From this, it follows that it will be possible to rewrite the differential $f(x, y)dx$ as

$$f(x, y)dx = \phi x.da + \phi_1 x.da' + \phi_2 x.da'' + \cdots,$$

where $\phi x, \phi_1 x, \ldots$ are rational functions of $x, a, a', a'', \ldots$ Integrating, one obtains

$$\psi x = \int (\phi x.da + \phi_1 x.da' + \cdots)$$

Notice that this equation remains valid when one replaces x by the μ values of this quantity. From this we deduce:

$$\psi x_1 + \psi x_2 + \cdots + \psi x_\mu$$

$$= \int [(\phi x_1 + \phi x_2 + \cdots + \phi x_\mu)da$$

$$+ (\phi_1 x_1 + \phi_1 x_2 + \cdots + \phi_1 x_\mu)da' + \cdots]. \qquad (10.7)$$

The coefficients of the differentials $da, da', \ldots$ in this equation are rational functions of $a, a', a'', \ldots$ and of $x_1, x_2, \ldots, x_\mu$. Moreover, they are symmetric with respect to $x_1, x_2, \ldots, x_\mu$. Therefore, according to a known theorem, it will be possible to express those functions rationally in terms of $a, a', a'', \ldots$ and the coefficients of the equation $\rho = 0$. But those coefficients are themselves rational functions of the variables $a, a', a'', \ldots$, therefore this will also be the case for the coefficients of $da, da', da'', \ldots$ in the Eq. (10.7). Consequently, by integration, one will have an equation of the form (10.6).

I hope that the reader who takes the time to understand it will agree with me that this is a beautiful proof.[1]

---

[1] In the French original, I made here an untranslatable pun: "l'Abel preuve!".

Chapter 11
Abel's Motivations

Abel's works, which we mentioned in the last two chapters, are the first in which the modern notion of genus appeared. Nevertheless, it was in a hidden form, a little like a secondary character remaining in the shadows, devoid of name. It is worth trying to understand better the general problems which preoccupied Abel at that time, of which [128] is only an expression. Happily for us, Abel wrote about these problems in an 1828 letter [2] to Legendre:

> Besides the elliptic functions, there are two other branches of analysis I dealt with, namely, the integration of algebraic differential formulas and the theory of equations. Using a particular method, I found many new results, which above all are very general. I started from the following problem of the theory of integration:
>
> "Given an arbitrary number of integrals $\int y\,dx$, $\int y_1 dx$, $\int y_2 dx$ etc., where $y, y_1, y_2, \ldots$ are arbitrary algebraic functions of x, find all possible relations between them which are expressible by algebraic and logarithmic functions".
>
> First I discovered that an arbitrary relation must be of the following form:
>
> $$A \int y\,dx + A_1 \int y_1 dx + A_2 \int y_2 dx + \cdots$$
> $$= u + B_1 \log v_1 + B_2 \log v_2 + \cdots$$
>
> where $A, A_1, A_2, \ldots, B_1, B_2, \ldots$ etc. are constants, and $u, v_1, v_2, \ldots$ are *algebraic* functions of x. This theorem makes the solution to the problem much easier; but the most important is the following:
>
> "If an integral $\int y\,dx$, where y is related to x by an arbitrary algebraic equation, may be *explicitly or implicitly* expressed using algebraic and logarithmic functions, then one may always assume that:
>
> $$\int y\,dx = u + A_1 \log v_1 + A_2 \log v_2 + \cdots + A_m \log v_m,$$
>
> where $A_1, A_2, \ldots$ are constants, and $u, v_1, v_2, \ldots, v_m$ are *rational functions* of x and y."

© Springer International Publishing Switzerland 2016
P. Popescu-Pampu, *What is the Genus?*, Lecture Notes in Mathematics 2162,
DOI 10.1007/978-3-319-42312-8_11

We see that Abel had the very ambitious program of finding *all* the relations which may exist between an arbitrary number of abelian integrals. This makes it possible to put on a common footing the various themes treated in [1]. The reader may find a detailed presentation of those themes in Kleiman's paper [115].

We will come back to some of the theorems discovered by Abel in Chap. 17.

Chapter 12
Cauchy and the Integration Paths

In Chap. 9, we saw the genus timidly appearing in Abel's investigations concerning so-called *abelian integrals*. Let us concentrate now on the definition of such an integral.

As shown by the proof presented in Chap. 10, Abel's writings are extremely algebraic. It is not clear whether he was working only with real numbers, with complex numbers, or instead formally, adjoining roots of equations each time that he needed them, without being preoccupied about their embedding in the set of complex numbers. Neither was he preoccupied by the interpretation of the integral (9.1). Perhaps he worked with it intuitively, as one worked with square roots of negative numbers before the existence of constructive interpretations of them.

In the same period, Cauchy was engaged in the development of the theory of *definite* integrals between two complex numbers. In the beginning (see for instance his 1825 paper [36]), his approach was rather computational. Later on, he developed a much more geometric vision, in which to *integrate* meant *to walk along paths drawn in the plane*.

It was at that time that the interpretation of complex numbers as points of a plane endowed with a cartesian coordinate system propagated. In this context, mathematicians tried to reinterpret the theories in which such numbers were involved using this geometric viewpoint. One of Cauchy's great contributions is to have accomplished this for the integral calculus. For instance, here is what he wrote about abelian integrals in his 1846 paper [37]:

> Till now, when I considered definite integrals which are relative to the various points of a closed curve described by a moving point whose rectangular coordinates represent the real part of an imaginary variable x and the coefficient of $\sqrt{-1}$ in this variable, I had supposed that, in each integral, the function under the $\int$ sign took back precisely the same value when, after having covered the whole curve, one came back to the starting point. But nothing prevents us from admitting that, in such an integral, the function under the sign $\int$, constrained if we wish to vary with x by imperceptible degrees, gets nevertheless various values at various times when the value of x comes back to the initial value. It is what will happen, in particular, if the function under the sign $\int$, constrained to vary by

© Springer International Publishing Switzerland 2016
P. Popescu-Pampu, *What is the Genus?*, Lecture Notes in Mathematics 2162,
DOI 10.1007/978-3-319-42312-8_12

imperceptible degrees with the position of the mobile point which we consider, encloses roots of algebraic or transcendental equations. Then, if the mobile point describes several times the same curve, the roots enclosed in the function under scrutiny may vary with the number of revolutions which will bring the mobile point back to its primitive position O, in such a way that a root of a given equation may be replaced, after one revolution is accomplished, by another root of the same equation. [...] But this will not prevent the integral from acquiring, after only one revolution of the mobile point, a determined value; and what is worth noting, is that this value will usually depend on the position of the mobile point but will be independent, under certain conditions, of the curve's shape.

What is involved here is that an "algebraic function" $y(x)$ is not a function in the modern set-theoretic interpretation, as it does not associate to an element of the domain *a single* element of the target, but *several*. At that time, one said that such a function was *multivalued* or *multiform*.

Let us be more precise. By definition, $y(x)$ satisfies an irreducible polynomial equation with complex coefficients $P(x, y) = 0$. One says that this relation "*defines y as an algebraic function of x*". Let us denote by n the degree in y of the polynomial $P(x, y)$. For a fixed value $x_0 \in \mathbb{C}$, one gets a polynomial equation in y. By the fundamental theorem of algebra, this equation admits as many roots, counted with multiplicities, as the degree in y of $P(x_0, y)$. For special values of $x_0 \in \mathbb{C}$, which we will call the *critical points* of $y(x)$, one has either:

- the degree of $P(x_0, y)$ is less than the degree n in the variable y of $P(x, y)$ (at least one of the roots of $P(x_1, y) = 0$ escapes to infinity when x_1 converges to x_0, that is, the curve with equation $P(x, y) = 0$ has the line with equation $x = x_0$ as an asymptote); or:
- certain roots of the equation come together to give a multiple root.

Therefore, the "algebraic function" $y(x)$ associates n distinct values of y to each value x outside the finite subset K of $\mathbb{C}$ of its critical points. We say that K is the *critical set* of the algebraic function.

For instance, Eq. (5.2) of the lemniscate defines the algebraic function $y(x)$ which may be expressed by radicals in the following way:

$$y = \pm\sqrt{\frac{-(2x^2 + 1) \pm \sqrt{8x^2 + 1}}{2}}.$$

This algebraic function does not admit vertical asymptotes, and its critical set is:

$$\left\{-1, 0, 1, \pm\sqrt{-\frac{1}{8}}\right\}.$$

But, as shown by Abel and Galois, in general there are no such "explicit" expressions, in radicals, for arbitrary algebraic functions. Therefore the only general way to concretely give an algebraic function is via its defining polynomial $P(x, y)$, which is assumed to be irreducible.

Let us come back to the abelian integrals. *How can one integrate an algebraic function that takes several values at each point?* And if one wants to assign a single value at each point, how are those values to be chosen at different points? Cauchy understood that there is no way to do this without supplementary conventions. If instead, the two limits of integration as well as the value of the function at the starting limit point are fixed, then one may make a *continuous* choice of values *by analytic continuation* along any *path* which goes from one limit point to the other one without passing through the critical points.

More precisely, the constraint of continuity allows one to define little by little a *univalued* function along the path, then to integrate it in the way Cauchy had defined earlier. A choice of *continuous* univalued function $y(x)$ on a region $U \subset \mathbb{C} \setminus K$, such that $P(x, y(x)) = 0$, is called a *determination* of the algebraic function defined by the equation $P(x, y) = 0$. When U is a disk which does not meet the critical set K, such a determination exists uniquely as soon as one chooses its value $y(x_0)$ at a point $x_0 \in U$ among the roots of $P(x_0, y) = 0$. The analytic continuation along a path c may then be done by covering the path with a finite number of disks which are small enough to avoid K, then by extending the determinations little by little, from each disk to the next one.

This definition of integration along a path can be thought of as a generalization of the notion of a *definite integral between two points* of $\mathbb{R}$. Cauchy discovered that the integral between two points of $\mathbb{C}$ (the "limites imaginaires" in the titles of [36, 37]) depends mildly on the chosen path: it remains unchanged as long as one deforms this path continuously, keeping its extremities fixed and avoiding the critical points. This was the starting point of the theory of *homotopy* (see Vanden Eynde's paper [175]).

The fact that $\int_{c_1} y \, dx = \int_{c_2} y \, dx$ if y is an algebraic function of x and the paths γ_1, γ_2 are homotopic with fixed end-points in $\mathbb{C} \setminus K$ comes from the fact that the differential form $y(x) \, dx$ is *closed*[1] on a region on which one chooses a continuous determination of $y(x)$, which is equivalent to the fact that the real and imaginary parts y_1, y_2 of y satisfy the so-called "*Cauchy–Riemann*" system of partial differential equations (see Chap. 42).

Let us explain this in modern terms. To give a homotopy between the two paths c_1, c_2 amounts to giving a differentiable map ϕ from a rectangle $ABCD$ to $\mathbb{C} \setminus K$, which sends the side AB to the starting point of the path, the side CD to the terminal point, the side BC to c_1 and AD to c_2 (see Fig. 12.1).

It is possible that the image of ϕ surrounds at least one of the critical points (as in our drawing) and that no determination of $y(x)$ exists on this image. Nevertheless, what counts is the rectangle, which contains no problematic points. Then, one takes the preimage $\phi^*(y \, dx)$ of the multivalued differential form on this rectangle. There, one may choose a unique determination of $\phi^* y(x)$, given its value at the starting

---

[1] The birth of the theory of differential forms in arbitrary dimensions is treated in Chap. 40.

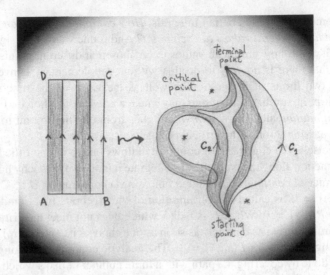

Fig. 12.1 A homotopy avoiding the critical points

point (therefore on the segment AB), and $\phi^*(y\,dx)$ becomes a closed univalued differential form. By Stokes' formula (see Theorem 41.1 below), the integral of $\phi^*(y\,dx)$ along the boundary of the rectangle vanishes. But the constancy of ϕ on each side AB and CD implies the vanishing of the integral on those sides, therefore: $\int_{AD}\phi^*(y\,dx) = \int_{BC}\phi^*(y\,dx)$, which is equivalent to the desired identity.

Chapter 13
Puiseux and the Permutations of Roots

As was explained by Cauchy in the excerpt of the paper [37] discussed in the previous chapter, if one takes a path which comes back to its starting point $\alpha \in \mathbb{C} \setminus K$ (that is, a *loop*), while avoiding the critical points of the algebraic function $y(x)$, one may get a determination of $y(x)$ which is different from the starting one. In fact, a little thought showed that to any loop based at α is associated in this way a permutation of the roots of the equation $P(\alpha, y) = 0$.

It was Puiseux who, in his 1850 paper [157], started the systematic study of the group of permutations obtained by considering all the loops based at a fixed point α. Thus, he understood that it is enough to study what happens when one encircles only one critical point. The reason is that every loop is homotopic to a composite of *elementary loops*, which encircle only one of those points (see the diagram 12 in Fig. 13.1 below, extracted from his article). Moreover, in [157, Chap. 30] Puiseux stated that this decomposition may be done in a single way if, in the sequence of elementary loops to be composed, a loop (called a "contour" by Puiseux) is never immediately followed by its inverse:

> [...] a closed contour of any shape passing through the point C (excluding the case where the contour passes through some of the points A, A', A'', etc.), which is covered in a determined sense, may be represented by the sequence of elementary contours to which one may reduce it by deformation: this sequence, all of whose terms must be endowed with a convenient sign, as we have explained before, is what we will call the *characteristic of the contour*. For instance, the contours *CLMC* of the *fig.* [...] 15, 16 being assumed to be covered in the sense *CLMC*, will have the characteristics
>
> $$[\ldots], (+A)(+A')(+A''),$$
> $$(+A')(+A'')(-A')(-A)(+A'),$$
>
> respectively and, if covered in the opposite sense, their characteristics will be
>
> $$[\ldots], (-A'')(-A')(-A),$$
> $$(-A')(+A)(+A')(-A'')(-A').$$

© Springer International Publishing Switzerland 2016
P. Popescu-Pampu, *What is the Genus?*, Lecture Notes in Mathematics 2162,
DOI 10.1007/978-3-319-42312-8_13

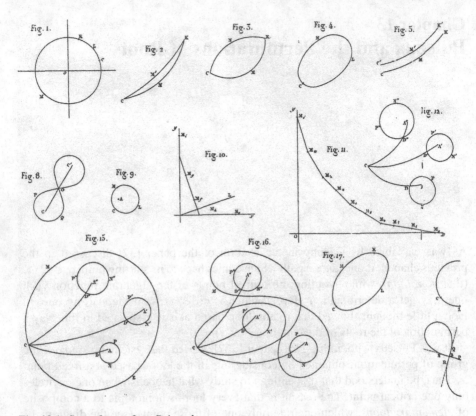

Fig. 13.1 Some figures from Puiseux's paper

[...] One sees easily that if the points A, A', A'', etc., remain the same, as well as the point C and the lines $CDA, CD'A', CD''A''$, etc., the same contour, covered in a fixed sense, may be given a single characteristic [...]. Two contours which have the same characteristic may always be reduced to each other without passing through any of the points A, A', A'', etc.; and, conversely, two contours which, for any choices of covering senses have distinct characteristics, cannot be reduced to each other.

In modern terms, Puiseux stated that *the fundamental group of the plane with a finite set of punctures is freely generated by the elementary loops*. As will be seen in Chap. 38, it was Poincaré who formulated the general notion of the *fundamental group*, partly with the aim of studying the permutations of the values of a multivalued function defined on an arbitrary space.

In order to understand how the roots are permuted when one covers an elementary loop, Puiseux used series expansions with fractionary exponents, objects which had been introduced by Newton. More precisely, let us assume that x_0 is a critical point and that y_0 is a multiple root of the equation $P(x_0, y) = 0$ (the case where a root escapes to infinity may be treated in an analogous way by making the change of variables $z = 1/y$). By taking as new variables $x - x_0$ and $y - y_0$, one may assume that $x_0 = y_0 = 0$.

Let us consider, for instance, the case of the lemniscate (5.2). The origin $(0, 0)$ is such a critical point (x_0, y_0). One sees that two *branches* meet at the origin, but that they become identical farther away. This corresponds to the fact that the defining polynomial, which is irreducible in the ring of polynomials $\mathbb{C}[x, y]$, becomes reducible in the ring of convergent power series $\mathbb{C}\{x, y\}$ in the variables x and y.

In general, the *branches* (or local irreducible components) at the point $(0, 0)$ of an algebraic plane curve C defined by the polynomial equation $P(x, y) = 0$, without constant term, are the loci of points defined in a sufficiently small neighborhood of $(0, 0)$ by the irreducible factors of $P(x, y)$ in the ring $\mathbb{C}\{x, y\}$.

Assume that $f(x, y) \in \mathbb{C}[[x, y]]$ defines a branch of C at the origin, and assume also[1] that f is polynomial in y. Denote by p its degree. Newton showed (see the extracts of the letters of Newton to Oldenburg presented in [25, pages 372–375]) that in this case there exists a power series $\lambda(t) \in \mathbb{C}[[t]]$ such that $\lambda(x^{1/p})$ is formally a root of the equation $f(x, y) = 0$. Because Puiseux needed to look at values of those *fractionary power series* $\lambda(x^{1/p})$ outside $x = 0$, he inquired about their convergence. He proved that they were indeed convergent in a neighborhood of 0.

Nowadays, such power series with rational exponents and bounded denominators are known as *Puiseux series* or as *Newton–Puiseux series*. The reader who wants to learn more about them, for instance how they are used in the study of singularities of plane algebraic curves, may consult the textbooks of Brieskorn and Knörrer [25], Fischer [77] or Wall [182].

Using such expansions in fractionary power series, Puiseux expressed each permutation of the roots obtained by turning along an elementary loop as a product of cycles. As explained before, this allowed him to deduce also an expression of the permutation which corresponds to any loop.

Note that the group of permutations obtained in this way is a *measure of the multivaluedness* of the algebraic function under scrutiny. Nowadays it is called its *monodromy group*. We won't pursue the development of this notion. Instead, in the next chapter we will pass to a completely different way of dealing with the phenomenon of multivaluedness, which was initiated by Riemann.

---

[1]This may always be arranged by the so-called *Weierstrass preparation theorem* (see [25]).

Chapter 14
Riemann and the Cutting of Surfaces

Shortly after Cauchy and Puiseux, Riemann came up with a radically different solution to the problem of multivaluedness of functions.

He started with the same thought experiment as Cauchy, that of *analytic continuation* along a loop. What changed is that he imagined a very thin "*sheet*" propagating along the loop and such that, when one comes back to the starting point, one arrives *on a different sheet* whenever the value of the function obtained by analytic continuation is different from its initial value.

In this way, by performing the analytic continuation along all the possible loops, one associates to the given algebraic function *a many sheeted smooth compact surface* which covers the *Riemann sphere* $\overline{\mathbb{C}} = \mathbb{C} \cup \{\infty\}$ associated with the plane $\mathbb{C}$ of one complex variable x. This surface was later called "*the Riemann surface associated with the multivalued function $y(x)$*".

Riemann introduced this construction as early as 1851, in [160], but it was in his 1857 paper [161, Preliminaries, I] where he described it in the most visual way[1]:

> In several investigations, for instance in the study of algebraic and abelian functions, it will be useful to represent the way in which a multivalued function ramifies, in the following geometric way:
>
> Conceive a surface extended along the (x, y) plane and coinciding with it (or, if one wants, an infinitely thin body covering the plane), which extends only so far as the function does. When the function extends, this surface will also extend with it. In a region of the plane where two or more extensions of the function occur, the surface will be double or multiple. It will consist of two or several sheets, each one of them corresponding to a branch of the function. Around a ramification point of the function, a sheet of the surface will extend to another sheet, and in such a way that, in the neighborhood of this point, the surface may be imagined as a helicoid whose axis is perpendicular to the (x, y) plane at that point, and whose thread is infinitely small. But when, after several turns of z [$= x + iy$] around the ramification value, the function takes back its initial value, one must assume that the superior sheet of the surface connects to the inferior one by traversing the rest of the sheets.

---

[1]In the next fragment, x and y denote the real and imaginary parts of the variable $z \in \mathbb{C}$.

© Springer International Publishing Switzerland 2016
P. Popescu-Pampu, *What is the Genus?*, Lecture Notes in Mathematics 2162,
DOI 10.1007/978-3-319-42312-8_14

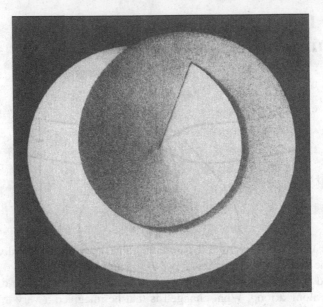

Fig. 14.1 The self-intersection has to be neglected!

    At each point of a surface which represents the way it ramifies, the multivalued function admits *a single* determined value, and may therefore be looked upon as a perfectly determined function of the place (of a point) on this surface.

This last sentence is essential: therefore, *the function which was multivalued in terms of the variable z becomes univalued as a function defined on the associated surface.*

If one wants to effectively construct in our physical space a *helicoid* as described by Riemann, one is forced to make the sheets pass through each other. This is visible in the model reproduced in Fig. 14.1 of a neighborhood of a ramification point on a Riemann surface with two sheets.

This figure comes from Neumann's book [139] of 1865, which was one of the first textbooks on the theory of Riemann surfaces.[2] It is important to understand that such a line of self-intersection is an artefact of the 3-dimensional spatial representation, and that one has to disregard it.

Indeed, the Riemann surface T associated with the algebraic function $y(x)$ is *smooth*: it contains no singular points. Above a critical point $x_0 \in \overline{\mathbb{C}}$, there are as many points of the surface as irreducible factors of the polynomial $P(x, y)$ in $\mathbb{C}\{x - x_0\}\,[y]$, the ring of polynomials in y with coefficients which are convergent power series in the variable $x - x_0$. If one of those irreducible factors has a Newton–Puiseux series of the form $\lambda((x - x_0)^{1/p})$, then $t = (x - x_0)^{1/p}$ is a local coordinate on the surface.

---

[2]I took the picture from [172, page 284].

Each point of the surface T lies above a point of the Riemann sphere $\overline{\mathbb{C}}$. Therefore, one has a holomorphic map $f : T \to \overline{\mathbb{C}}$ which is a *ramified covering*, a local homeomorphism outside the ramification points. In Riemann's vision, a *ramification point* of f is a point of T situated on the axis of a helicoid. Such an axis is illustrated by the line in 3-dimensional space which projects to the center of Fig. 14.1. The images of the ramification points by the map f are contained in the critical locus of the given algebraic function $y(x)$. But not all critical points of $y(x)$ are images of ramification points, as shown again by the example of the lemniscate (see Eq. (5.2) and the accompanying Fig. 5.1). In that case, the origin of the x-axis is critical because a singular point of the curve projects onto it. This singular point corresponds to two points of the associated Riemann surface, neither of them being a ramification point.

The rational functions in the variables x, y become *meromorphic* on T. That is, in terms of a local coordinate t on that surface, they may be written as $t^m u(t)$, where $m \in \mathbb{Z}$ and $u(t)$ is a convergent power series which does not vanish at $t = 0$. The rational function has a zero of order m if $m > 0$ and a pole of order $-m$ if $m < 0$.

An abelian integral $\int y \, dx$, where y is an algebraic function of x, may be reinterpreted on the Riemann surface T associated with $y(x)$. More precisely, if one asks to compute it "à la Cauchy", along a path contained in the plane of the complex variable x and avoiding the critical set, the datum of the initial value of the function $y(x)$ determines the starting point on T. Then the path lifts canonically to T and one computes along that lift the value of the integral of $y \, dx$, which is a differential form of degree 1, univalued on T. The specificity of this form is that it is *closed*: $d(y \, dx) = 0$.[3]

Riemann was brought in this way to study the integrals of closed differential forms of degree 1 along paths drawn on a surface T. Here is what he wrote on the subject in [161, Preliminaries, II][4]:

Consequently, the integral

$$\int (Xdx + Ydy),$$

taken between two fixed points, along two distinct paths, has the same value when the union of those two paths forms the whole boundary of part of the surface T. It follows that, if any curve in the interior of T, which closes up by coming back onto itself, forms the complete boundary of a part of T, then the integral between two fixed points has always the same value which is a function of the position of the final point everywhere on T and is independent of the path of integration. [...]

Whenever it is possible to draw n closed curves $a_1, a_2, \ldots, a_n$ on a surface T, which, either taken separately, or taken together, do not form the complete boundary of part of the surface, but which, joined to any other closed curve, form the boundary of part of the surface, then the surface will be called $(n + 1)$-ply connected. [...]

---

[3]In Chap. 40 we will see how one arrived, historically speaking, at this mode of expression.

[4]In this text, x and y again denote the real and imaginary parts of a complex variable z, which is a local complex coordinate on the surface T.

An $(n + 1)$-ply connected surface is decomposed [...] into an n-ply connected one by any cross-cut which does not decompose it into separate pieces.

In order to make those considerations applicable to a surface without boundary, that is, to a closed surface, it will be necessary to turn this closed surface into a surface with boundary, by making a puncture at any point, such that the first decomposition will be done using a cross-section which starts from this point and returns to it, and which forms consequently a closed curve.

Riemann defined therefore a complexity for compact surfaces which possibly have boundaries, namely, their *"connection order"*. A sphere has connection order 1 and a torus 3. For a surface without boundary, this connection order may be expressed simply in terms of its genus: it is its double plus one.

In fact, here is the way *the genus*, which Riemann denoted by p, appeared in his paper (see [161, Chaps. I; III]):

Conceive now that is given a connected surface T, covering everywhere n times the z-plane, without boundary, but which may be looked upon, by the previous considerations, as a closed surface, and that this surface was decomposed into a simply connected surface T'. As the boundary curve of a simply connected surface is formed by a unique contour, a closed surface takes an even number of boundary components when using an odd number of cross-sections and an odd number of such boundary components when using an even number of such sections. In order to carry out this decomposition of the surface, it will therefore be necessary to make an even number of sections. Let $2p$ be the number of those cross-sections.

In other words, $2p$ is the maximum number of loops[5] which may be drawn on the surface T without disconnecting it. It is at this point that the genus became a *topological attribute of a surface*, instead of an *algebraic attribute of a multivalued function*, as in Abel's work.

We will see in Chaps. 38 and 39 that, following the works of Poincaré, this number $2p$ was to be reinterpreted as the rank of the first *homology group* of T, that is, as its first *Betti number*. In Theorems 39.2 and 42.2 are stated two generalizations in higher dimensions of the fact that the first Betti number of an orientable surface is *even*.

Let us come back to the spirit of Riemann's construction. It is possible that on the surface T, the integral of the differential form under scrutiny may be multivalued. By Stokes' formula,[6] one sees that this is caused by the presence of curves which do not bound portions of the surface. Riemann then artificially created *barriers*, until each closed curve which avoids them bounds a portion of the surface, and he cut along them.

It is a specificity of dimension 2 that the surface obtained by this cutting process is automatically simply connected (it is in fact homeomorphic to a disk). Indeed, in his paper [151] published in 1904, Poincaré gave the first example of a manifold of

---

[5]Note that here, in contrast to the situation considered in the Introduction, the loops are allowed to intersect.

[6]See Theorem 41.1 for its general modern statement. Here, in the context of curvilinear integrals on surfaces, it is often called the *Green–Riemann formula*.

dimension 3 which has the same homology as the 3-dimensional sphere, but which is not simply connected. This made Poincaré ask:

Is every 3-dimensional simply connected closed manifold homeomorphic to a sphere?

This question became known as *the Poincaré conjecture*, one of the most famous conjectures of the twentieth century. It was proved only almost 100 years later, in [146], by Perelman. The reader eager to learn more about 3-dimensional topology, a domain we will touch again only in Chap. 19, may read Milnor's articles [133, 134].

Let us come back to Poincaré's 3-dimensional example. Removing a ball from it, one obtains a manifold bounded by a 2-dimensional sphere, in which any closed curve or surface necessarily bounds, but which is not simply connected. This contrasts with the 2-dimensional situation.

It is possible that Riemann heard from Gauss the suggestion to analyze the surfaces by successive cross-sections. An indication in this sense is given to us by a letter of Betti to his friend Tardy, dated 6 October 1863, and republished in [189][7]:

> What gave Riemann the idea of the cuts was that Gauss defined them to him, talking about other matters, in a private conversation. In his writings one finds that analysis situs, that is, this consideration of quantities independently from their measure, is "wichtig" [important].

At this point, a definition of the genus of an algebraic function $y(x)$ in the spirit of "analysis situs" (that is, a topological definition) has been made available. But how can it be used in order to compute the value of the genus for concrete algebraic functions?

Consider the ramified cover $f : T \to \overline{\mathbb{C}}$ associated with the algebraic function $y(x)$. If P is a ramification point, and if f is given in convenient local coordinates centered at the points P and $f(P)$ by the map $z \to z^m$, one says that the *ramification index* $i(P)$ of f at the point P of T is $m - 1$. This vocabulary may be extended to the case of a ramified cover between two Riemann surfaces. One then has the following so-called *Riemann–Hurwitz formula* relating the genera of the two surfaces, the number of sheets and the indices of ramification of f:

Theorem 14.1 *Let $f : T \to S$ be an n-sheeted ramified cover between two compact Riemann surfaces. Then:*

$$p(T) = n \cdot p(S) + \sum_{P \in T} i(P).$$

The simplest way to prove this theorem is probably to consider a triangulation of S whose vertices contain the images of the ramification points of f, then to lift it to a triangulation of T. One then expresses the Euler–Poincaré characteristic of T in terms of that of S, of the number of sheets (called the *degree* of the covering)

---

[7]The translation into English was done by Weil.

and of the ramification indices. Finally, one uses the fact that the Euler–Poincaré characteristic of a Riemann surface of genus p is $2 - 2p$ (see Theorem 47.3).

Such a proof was apparently imagined only after Poincaré had established a notion of Euler–Poincaré characteristic for manifolds of arbitrary dimension (see Chap. 47).

One may find in Pont's book [152] many more details about the topological works of Gauss, Riemann, and more generally of the mathematicians of the nineteenth century. There, one may discover the essential influence of Gauss on the first directions of research in this domain. For instance (see also [153]), it was Listing, one of Gauss' students, who invented the term *"topology"*, which was to replace in the twentieth century that of *"analysis situs"*, proposed by Leibniz.

Chapter 15
Riemann and the Birational Invariance of Genus

As explained in the previous chapter, although Riemann developed a very topological vision of surfaces, he did not forget that his fundamental examples came from algebraic functions and their integrals.

Changes of variables are essential for integral calculus. It is natural to expect that certain of them preserve interesting properties of the studied integrals. It was probably considerations of this kind that led Riemann to identify in [161, I ; XII] polynomial equations which are related by rational transforms:

> We will now consider, as parts of a *single class all the irreducible algebraic equations between two variables which may be transformed into one another by rational substitutions*; in this way,
>
> $$F(s, z) = 0 \text{ and } F_1(s_1, z_1) = 0$$
>
> belong to the same class when one may replace s and z by rational functions of s_1 and z_1, such that $F(s, z) = 0$ gets transformed into $F_1(s_1, z_1) = 0$, s_1 and z_1 equally being rational functions of s and z.

In geometric terms, the plane algebraic curves defined by the equations $F(s, z) = 0$ and $F_1(s_1, z_1) = 0$ are called *birationally equivalent*. In a modern formulation, an essential theorem of Riemann is:

Theorem 15.1 *The genus of an algebraic curve is a birational invariant.*

This theorem may be seen as a consequence of the fact that a birational equivalence induces a homeomorphism between the associated Riemann surfaces, and that the genus is invariant under homeomorphisms (see Chap. 18). This fact is specific to complex dimension 1. Indeed, due to the appearance of the phenomenon of *blowing up* (see Chap. 29), starting from complex dimension 2 one finds an infinite number of topological types in any class of birationally equivalent smooth algebraic varieties.

© Springer International Publishing Switzerland 2016   41
P. Popescu-Pampu, *What is the Genus?*, Lecture Notes in Mathematics 2162,
DOI 10.1007/978-3-319-42312-8_15

Coming back to the objects manipulated so far, note the way in which the attention moved from:

- an *abelian integral* $\int y\,dx$; to
- the *algebraic function* $y(x)$; to
- the *polynomial* $P(x, y)$ which defines it; to
- the *plane curve* C defined by the vanishing of the polynomial; and finally to
- the associated *Riemann surface*, on which the differential form $y\,dx$ and the algebraic function $y(x)$ become single-valued.

In fact, the latter two algebraic objects $y\,dx$ and $y(x)$ also become single-valued on the curve C. But for Riemann it was important to have a *smooth* and compact domain, a reason for which it was not satisfactory to work with C. Of course, at his time it was already known how to compactify (a concept which existed only intuitively) by passing from the curves in the affine plane to their closures in the projective plane. This process adds "*points at infinity*" but, by contrast with Riemann's construction, it does not remove the singular points.

In Chap. 29 we will see that this requirement of passing from an arbitrary algebraic variety to a smooth one would bring serious problems in higher dimensions.

Chapter 16
The Riemann–Roch Theorem

Any multivalued algebraic function $y(x)$ becomes univalued on the Riemann surface T associated with it. Consequently, if $Q(x, y)$ is a rational function with two variables, then the algebraic function $Q(x, y(x))$ also becomes univalued on T. The functions of this type form *the field of rational functions on the Riemann surface*, which is the fundamental algebraic object associated with T.

A rational function of one variable is determined, up to a multiplicative constant, by its zeros and poles, counted with multiplicities. This property remains true on any Riemann surface. One may prove it by applying the *maximum principle* to the everywhere holomorphic quotient of two functions satisfying the previous constraints. By this principle, the absolute value of a holomorphic function has a local maximum only if the function is constant.

What does not remain true in general, being a specificity of genus 0, is that *any finite set of points on T, endowed with integer multiplicities with vanishing sum, is the locus of zeros and poles of some meromorphic function.*

In order to see that this property is no longer true in genus 1 or higher, it is enough to remark that given two distinct points A and B on T, a meromorphic function which has a single zero at A and a single pole at B is necessarily of degree 1 when seen as a map $T \to \overline{\mathbb{C}}$. This implies that it realizes an isomorphism between T and the Riemann sphere. Therefore T is necessarily of genus 0. Conversely, if T is of genus 0, then it is isomorphic to $\overline{\mathbb{C}}$ (see Theorem 16.2 below).

Riemann understood that in general it was more fruitful not to specify *all* the zeros and poles, but only a set of $n \geq 0$ points, including the allowed poles, without constraining the zeros. On $\overline{\mathbb{C}}$, the dimension of the complex vector space of meromorphic functions whose polar locus is contained in this fixed set is always $n + 1$. In general, the genus p reduces this number. More precisely, Riemann showed in his 1857 paper [161] that:

Theorem 16.1 *Let T be a compact Riemann surface of genus $p \geq 0$ with n distinct marked points. Then, the dimension of the complex vector space of meromorphic*

© Springer International Publishing Switzerland 2016
P. Popescu-Pampu, *What is the Genus?*, Lecture Notes in Mathematics 2162,
DOI 10.1007/978-3-319-42312-8_16

functions on T with simple poles, all of them contained among the marked points, is at least $1 - p + n$.

In particular, by taking $p = 0$ and $n = 1$, we see that on any Riemann surface T of genus 0, there exists a meromorphic function with a unique and simple pole. This function realizes an isomorphism from T to $\overline{\mathbb{C}}$. Therefore:

Theorem 16.2 *Up to isomorphism, there exists only one Riemann surface of genus 0, which is the Riemann sphere* $\overline{\mathbb{C}}$.

The algebraic curves whose Riemann surface has genus 0 may therefore be parametrized birationally using rational functions. For this reason, they are called *rational curves*.

In his 1865 paper [162], Riemann's student Roch succeeded in explaining the reason why equality need not always hold in the previous theorem:

Theorem 16.3 *Let T be a compact Riemann surface of genus $p \geq 0$ with n distinct marked points. Then, the dimension of the complex vector space of meromorphic functions on T with simple poles, all of them contained among the marked points, is equal to the sum of the expression $1 - p + n$ and of the dimension of the complex vector space of holomorphic 1-forms on T which vanish at the marked points.*

We see that the differential forms play a role in this question, even if they do not appear explicitly in the statement of the initial problem.

Note that this "Riemann–Roch theorem", which also holds when one interprets "Riemann surface" in the abstract sense of a smooth compact holomorphic curve (see Chap. 23), shows that there exist plenty of non-constant meromorphic functions on them. This implies with a little more work that *any smooth compact holomorphic curve embeds in a projective space.*

On the other hand, starting from complex dimension 2, there exist smooth compact holomorphic manifolds which cannot be embedded in any projective space (one says in this case that they are not projective). We will return to this phenomenon at the end of Chap. 42.

The Riemann–Roch theorem became a powerful guide for the extension of the notion of genus in higher dimensions: such an extension should serve to formulate a generalization of the theorem. We will discuss generalized Riemann–Roch theorems in Chaps. 44, 49 and 50.

Chapter 17
A Reinterpretation of Abel's Works

We saw in Chap. 11 that Abel had a very ambitious program aiming to study all possible relations between abelian integrals. Even if he did not always completely prove them, he discovered many theorems concerning those relations. For instance, in [115] Kleiman lists and sketches modern proofs of the theorems he encountered in Abel's paper [1]. One of the most famous such theorems, which is explained in nearly every textbook on algebraic curves and Riemann surfaces, is the following one:

Theorem 17.1 *Let T be a compact Riemann surface. Let us consider two sets* $\{A_i\}_i$ *and* $\{B_j\}_j$ *of n points on it, and for every point* A_i, *a path* c_i *joining* A_i *to one of the points* B_j, *establishing in this way a bijection between the two sets of points. Then the sum* $\sum_j \int_{c_j} \omega$ *vanishes modulo periods of* ω, *for any holomorphic form* ω *of degree 1, if and only if there exists a meromorphic function on T whose only poles are the points* B_j *and only zeros are the points* A_i *(one then says that the system of points* A_j *is* linearly equivalent *to the system of points* B_j*).*

We saw in Chap. 12 that the definite integral of an algebraic function does not only depend on its end-points, but also on the homotopy class of the path which joins them in $\mathbb{C}$. With Riemann's viewpoint, this may be reformulated as the fact that an integral of a meromorphic form on a Riemann surface does not only depend on the end-points of the path of integration, but also on its homotopy class with fixed end-points. It is important then to understand how the value of the integral is modified when one changes the path.

The difference of the integrals computed on two paths with the same end-points is an integral on a closed curve. Such an integral is called a *period* of the form which is being integrated.

Let us come back to Theorem 17.1. Replace the paths c_j of its statement by other paths c_j' which still satisfy the same constraints. Taken together, the paths c_j followed by the paths c_j' with the opposite orientation form a finite set of closed curves drawn on the Riemann surface (see the example in Fig. 17.1). When one changes the paths

© Springer International Publishing Switzerland 2016
P. Popescu-Pampu, *What is the Genus?*, Lecture Notes in Mathematics 2162,
DOI 10.1007/978-3-319-42312-8_17

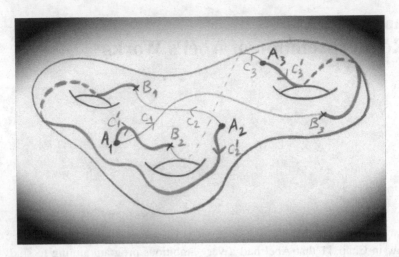

Fig. 17.1 Possible integration paths with fixed end-points

of integration, the sums considered in Theorem 17.1 are therefore modified by the addition of a *period* of ω. This fact explains the presence of those periods in the statement of the theorem.

It is worth noting that Abel proved only the necessity of the condition regarding the meromorphic function. He did not speak about periods but instead he worked formally with his integrals, as illustrated by the proof reproduced in Chap. 10. Nevertheless, he showed that if one considers the points B_j as variables, depending algebraically on a parameter t, that is, such that for every value of t they form the complete preimage $f^{-1}(t)$, where $f : T \to \overline{\mathbb{C}}$ is a meromorphic function, then the abelian sum:

$$\sum_i \int_{A_i}^{B_i(t)} \omega$$

is *constant*.

Let us give a geometric explanation of this necessary condition, using Riemann surfaces.

The meromorphic function f, with given fixed sets of zeros and poles, represents T as a ramified cover of the Riemann sphere $\overline{\mathbb{C}}$. Outside the ramification points, $\overline{\mathbb{C}}$ is locally covered by n sheets of T. The sum of the restrictions of ω to those n sheets yields a holomorphic form on $\overline{\mathbb{C}}$, defined outside the ramification points. A local study in their neighborhood shows that this form extends to an *everywhere holomorphic* form ϕ on $\overline{\mathbb{C}}$. But on the sphere, such a form is always zero! Indeed, one has the following theorem, which characterizes the genus in two ways by means of integral calculus (instead of holomorphic forms, Riemann still spoke, like Abel, of *integrals of the first kind*):

Theorem 17.2 *Let T be a compact Riemann surface of genus p. Then:*

- *the dimension of the complex vector space of the forms of degree 1 which are holomorphic on T is equal to p;*
- *the number of zeros, counted with multiplicities, of any non-identically zero holomorphic 1-form on T is equal to $2p - 2$.*

Now one may easily finish the argument. Indeed, by hypothesis, the points A_j form the preimage $f^{-1}(0)$ and the points B_j the preimage $f^{-1}(\infty)$. One may therefore choose as paths c_j the various preimages by f of a fixed path c joining 0 and ∞ in $\overline{\mathbb{C}}$, and which avoids the ramification points. The abelian sum taken along those paths is therefore equal to the integral $\int_c \phi$. As ϕ vanishes, we are finished.

A generalization of the previous implication, also proved by Abel (who, of course, always spoke of algebraic functions rather than Riemann surfaces) is the following:

Theorem 17.3 *Let T be a compact Riemann surface, and let ω be a meromorphic form of degree 1 on T. Assume that for any $t \in \overline{\mathbb{C}}$, the points $B_i(t)$ form the preimage $f^{-1}(t)$ of a meromorphic function $f : T \to \overline{\mathbb{C}}$. Then the abelian sum:*

$$\sum_i \int_{B_i(0)}^{B_i(t)} \omega$$

has the form $R(t) + \log Q(t)$, where $R(t)$ and $Q(t)$ are rational functions in the variable t.

The ambiguity in the choice of paths of integration manifests itself by the ambiguity in the choice of determination of logarithm.

The reader curious to learn about the numerous further developments of these ideas may read Griffiths' papers [86, 87].

Note that in Theorems 17.1 and 17.3, the paths of integration depend on a parameter. In fact, Abel also studied what happens when one takes such sums where the paths have independent ends:

Theorem 17.4 *Consider a compact Riemann surface T of genus p and a meromorphic form ω of degree 1 on T. Fix a point $A \in T$. Then, the abelian sum $\sum_{j=1}^{p+1} \int_A^{B_j} \omega$ is equal to the sum of an elementary function in the B_j's and an abelian sum $\sum_{k=1}^{p} \int_A^{C_k} \omega$, where the points C_k algebraically depend on the points B_j.*

The previous theorem is Abel's huge generalization of the addition Theorem 7.1 for the lemniscatic integral. As usual, Abel interpreted this result as a concrete theorem about functions in several variables. Using Riemann's reformulation, one gets a function defined on the $(p+1)$-th power of the Riemann surface: a manifold of dimension $(p+1)$. This is one of the reasons why Riemann's paper [161] stimulated not only research on smooth complex curves, that is, Riemann surfaces, but also on higher dimensional complex manifolds.

For the sake of the reader sufficiently accustomed to the manipulation of such manifolds, let us explain briefly how a higher dimensional geometric perspective

allows one to understand Abel's theorem stated in the introduction of [1], in which
we recognized (see Chap. 9) the first manifestation of the modern notion of genus.

Start from a compact Riemann surface T, for instance the one associated with an
algebraic function. Fix a starting point $A_0 \in T$ for the computation of the integrals
along paths drawn on T, as well as an arbitrary meromorphic 1-form ϕ on T. Abel
was interested in the function:

$$(A_1, \ldots, A_m) \to \sum_{j=1}^{m} \int_{c_j} \phi$$

where c_j is any path going from A_0 to A_j.

Fix a basis $(\omega_1, \ldots, \omega_p)$ of the vector space of holomorphic forms on T (see Theorem 17.2). To every loop c drawn on T, assign a *period-vector* $(\int_c \omega_1, \ldots, \int_c \omega_p) \in$
$\mathbb{C}^p$. Those various vectors form a subgroup Γ of $(\mathbb{C}^p, +)$, which is generated by the
period-vectors taken along a set of $2p$ loops which do not disconnect T. One shows
that those $2p$ vectors are independent, and that moreover Γ is *discrete* in $\mathbb{C}^p$. This
allows one to view the quotient space $\mathbb{C}^p / \Gamma$ as a topological torus of dimension $2p$.
One calls it the *Jacobian* of the Riemann surface T. Let us denote it by $J(T)$.

Jacobi's name is attached to this notion since he was the first to develop the
idea that one has to study *simultaneously* all the abelian integrals of the first kind
associated to an algebraic curve. His first paper on this subject is [107]. Incidentally,
it is in the same paper that he introduced the terminology "*abelian transcendents*"
for abelian integrals.

Consider the space T^m of m-tuples of points of T. Having fixed the starting point
A_0, one may associate to any point $(A_1, \ldots, A_m) \in T^m$ the vector of abelian sums:

$$(\sum_{j=1}^{m} \int_{c_j} \omega_1, \ldots, \sum_{j=1}^{m} \int_{c_j} \omega_p).$$

As this sum is well-defined modulo period-vectors, its image in the Jacobian $J(T)$
does not depend on the choice of paths of integration. This gives therefore a map:

$$\Phi_m : T^m \to J(T).$$

One may show that Φ_m is *surjective* as soon as $m \geq p$.

Let us assume that this inequality $m \geq p$ holds. By Theorem 17.1, the preimages
$\Phi_m^{-1}(t)$ form a linear equivalence class of *ordered* m-tuples on T. The surjectivity
of Φ_m shows that for all the points $t \in J(T)$ outside a certain strict complex
subvariety of $J(T)$, $\Phi_m^{-1}(t)$ has complex dimension $\dim T^m - \dim J(T) = m - p$.
Therefore, if one constrains the m-tuple $(A_1, \ldots, A_m) \in T^m$ to the preimage $\Phi_m^{-1}(t)$,
one establishes (locally in T^m) exactly p algebraic relations among the variables
$A_1, \ldots, A_m$ of the functions $A_i \to \sum_{j=1}^{m} \int_{c_j} \omega_i$. This explains why we asserted at the
end of Chap. 9 that the number of these relations is equal to the genus.

These relations are a priori transcendental (that is, holomorphic but not necessarily algebraic), because the Jacobian $J(T)$ is by construction only a *smooth complex analytic manifold*. In fact, and this is a deep theorem, the Jacobian may be holomorphically embedded into a projective space, which endows it with an induced structure of an algebraic manifold.[1] Using this algebraic structure, the map Φ_m becomes a morphism of algebraic varieties. Therefore, the relations which define one of the preimages $\Phi_m^{-1}(t)$ may be chosen to be algebraic, by taking the preimages under Φ_m of a set of rational functions on $J(T)$, whose vanishing locus is the point t. As expected, this approach agrees with Abel's assertion.

Let us come back to the meromorphic differential form ϕ on T. By taking its preimages under the m projections of the cartesian product T^m onto its factors T, and summing those preimages, one gets a closed meromorphic form of degree 1 on T^m. By construction, it is invariant under the permutations of the factors T of T^m. Therefore, it descends to the quotient $T^{[m]}$ of T^m by this group of permutations.

Denote by $\phi^{[m]}$ the resulting closed form of degree 1 on $T^{[m]}$. Note that the points of the space $T^{[m]}$ parametrize the unordered m-tuples of points of T, possibly coincident, that is, the *effective divisors* of degree m (see the Chap. 28).

The quotient of a preimage $\Phi_m^{-1}(t)$ by the previous group of permutations is equal to the space of effective divisors which are linearly equivalent to a fixed effective divisor D. Such a linear equivalence class is called a *complete linear series*. It is naturally a complex projective space, as it may be identified with the projectivization of the complex vector space of meromorphic functions which have poles at most along the fixed divisor D.

The preimage $\Phi_m^{-1}(t)$ is therefore naturally a complex projective space. On the other hand, one may show that a closed meromorphic form of degree 1 on a projective space of arbitrary dimension is always the differential of the sum of an algebraic function and of the logarithm of an algebraic function.

This explains geometrically the theorem stated by Abel in his introduction to [1], and the role played in it by the notion of genus.

The reader eager to learn much more about the developments of Abel's ideas until the present may consult the various articles of [127]. In order to be further acquainted with the relations between Riemann surfaces and their Jacobians, one may read Mumford's paper [138].

Let us come back now to the topological interpretation of the genus.

---

[1]One may read Catanese's paper [35] for learning about the relations of such algebraization theorems with Abel's works.

Chapter 18
Jordan and the Topological Classification

In his paper [109], which appeared in 1866, Jordan described one of the first attempts to prove a theorem on the topological classification of surfaces, that is, *up to homeomorphisms*.[1] The term *homeomorphism* did not yet exist and Jordan was one of the first people, with Mobius, to offer a definition of this notion. His definition appeals a lot to intuition:

> We will rely [...] on the following principle which may be regarded as obvious [...]:
> *Two surfaces S, S' are applicable on each other if one may decompose them into infinitely small elements, such that arbitrary contiguous elements of S correspond to contiguous elements of S'.*

Here is the way he presented his problem, as well as the theorem he obtained:

> One of the best known problems of Geometry is the following one:
> *Find the necessary and sufficient conditions for two surfaces or pieces of surfaces which are flexible and inextensible to be applicable on each other without tearing or duplication.*
>
> One may state an analogous problem, supposing instead that the surfaces are extensible in an arbitrary way. The question simplified like this belongs to the geometry of situation, and we will solve it by proving the following theorem:
>
> **Theorem** *In order for two surfaces or pieces of surfaces which are arbitrarily flexible and extensible to be applicable on each other without tearing or duplication, it is necessary and sufficient:*
>
> 1° *That the number of separate contours which bound respectively those two portions of surfaces be the same (if the surfaces under consideration are closed, that number is zero).*
> 2° *That the maximal number of closed contours without self-crossings or mutual crossings, which may be drawn on each surface without decomposing it into two separate regions, be the same.*

---

[1] Another attempt was described by Möbius in [135]. This paper was commented by Pont in [150]. Nowadays, the first rigorous proof from our modern perspective is generally considered to be that given by Radó [158].

© Springer International Publishing Switzerland 2016
P. Popescu-Pampu, *What is the Genus?*, Lecture Notes in Mathematics 2162,
DOI 10.1007/978-3-319-42312-8_18

In fact, Jordan implicitly considers here only *orientable* surfaces. It was Möbius who, almost simultaneously, brought attention to the necessity of differentiating carefully between orientable and non-orientable surfaces, giving in particular the famous example of his band (see [153]). Note that the Riemann surfaces are all orientable, as they are ramified covers of the Riemann sphere $\overline{\mathbb{C}}$, which is orientable.

If the surface considered by Jordan is closed, then *"the maximal number of closed contours without self-crossings or mutual crossings, which may be drawn on each surface without decomposing it into two separate regions"* is precisely one of the definitions of the genus of the surface. Namely, it is the one we spoke about in the Introduction, when we drew curves surrounding holes. Notice that, even if the surfaces Jordan thought about are located in ordinary space, he never wrote about visible *holes*. Apparently, one had to wait for Clifford in order to see this intuitive notion of *hole* being used in order to make the notion of genus more easily understandable to novices. The next chapter is dedicated to this viewpoint.

Chapter 19
Clifford and the Number of Holes

We saw that Riemann denoted the genus by "p", a notation which is still frequently used today, in particular for the generalizations of this notion in higher dimensions. On the other hand, Riemann did not give a name to this notion, and his definition was not the one we saw in the Introduction, in terms of holes. There is a good reason for this, namely, that the surfaces imagined by Riemann consisted of sheets which thinly cover the plane, and therefore do not admit visible holes.

It was Clifford who, a little later, had the idea of defining the genus of a surface by counting its holes. The reason being that he first proved that a Riemann surface is necessarily homeomorphic to a surface admitting holes, embedded in the usual 3-dimensional space, as in the examples of the first image of the Introduction. This is precisely stated in his paper [47] of 1877:

> The object of this Note is to assist students of the theory of complex functions, by proving the chief propositions about Riemann's surfaces in a concise and elementary manner. [...]
>
> If two variables s and z are connected by an equation [polynomial of degree n in s and m in z], each is said to be an algebraic function of the other. Regarding z as a complex quantity $x + iy$, we represent its value by the point whose co-ordinates are x, y, on a certain plane. To every point in this plane belongs one value of z, and consequently, in general, n values of s, which are the roots of the equation [...].
>
> We shall now go on to shew that this n-valued function, which we have spread out upon a single plane, may be represented as a *one*-valued function on a surface consisting of n infinite plane sheets, supposed to lie indefinitely near together, and to cross into one another along certain lines. [...]
>
> Let now this n-fold plane be inverted in regard to any point outside it, so that it becomes an n-fold sphere passing through the point. [...]
>
> We shall now prove that this n-fold spherical surface can be transformed without tearing into the surface of a body with p holes in it. [...]
>
> A closed curve drawn on a surface is called a *circuit*. If it is possible to move a circuit continuously on the surface until it shrinks up into a point, the circuit is called *reducible*; otherwise it is *irreducible*. In general there is a finite number of irreducible circuits on a closed surface which are *independent*, that is, no one of which can be made by continuous motion to coincide with a path made out of the others. [...]

© Springer International Publishing Switzerland 2016
P. Popescu-Pampu, *What is the Genus?*, Lecture Notes in Mathematics 2162,
DOI 10.1007/978-3-319-42312-8_19

[...] on the surface of a body having p holes through it, there are $2p$ independent irreducible circuits; one *around* each hole, and one *through* each hole.[1]

Another rather intuitive manner to define the genus of a closed orientable surface topologically is to decompose the surface as a *connected sum*.

In general, if S_1, S_2 are two oriented and connected surfaces, their *connected sum* $S_1 \sharp S_2$ is a new oriented surface, which is constructed by taking out a compact disk D_i from each surface S_i and by identifying the two resulting boundary circles by a diffeomorphism compatible with the orientations of the two surfaces $S_1 \setminus D_1$ and $S_2 \setminus D_2$. This is illustrated in Fig. 19.2.

In this way, one gets a composition law on the set of diffeomorphism classes of closed, oriented and connected surfaces. The diffeomorphism class of the spheres is a neutral element for this law. A surface S is called *prime* if it is not a sphere and if it cannot be written non-trivially as a connected sum (that is, if one writes $S = S_1 \sharp S_2$, then necessarily one of the surfaces S_1 and S_2 is a sphere). One shows that *there is only one prime orientable surface, the torus*.

Here is the announced interpretation of the genus:

Theorem 19.1 *If one decomposes an orientable closed surface of genus p as a connected sum of prime surfaces (that is, of tori), then there are exactly p of them.*

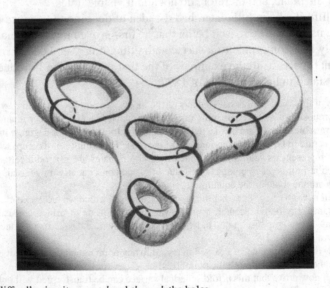

Fig. 19.1 Clifford's circuits *around* and *through* the holes

---

[1]The curves *around* each hole are for instance those represented in Fig. 4 of the Introduction. In Fig. 19.1 is also represented a curve *through* each hole.

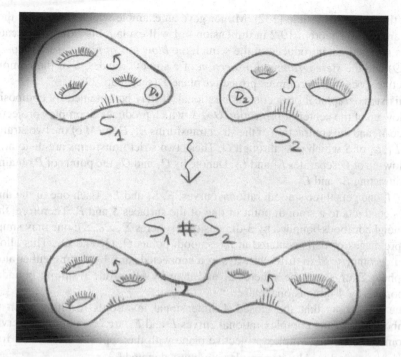

Fig. 19.2 The connected sum of two surfaces

An analogous vision was developed in the twentieth century for the oriented 3-dimensional closed manifolds. More precisely, the definition of *prime 3-manifolds* mimicking that of prime surfaces, one has:

Theorem 19.2 *Each closed, connected and oriented 3-dimensional manifold M may be written as a connected sum of a finite number of prime manifolds. Moreover, up to permutations, those are independent of the chosen decomposition.*

The first statement was proved in 1929 by Kneser in [119] and the second one in 1962 by Milnor in [132]. By analogy with Theorem 19.1, one could define the *genus* of a 3-dimensional manifold as the number of prime factors which appear in such a decomposition. But this denomination is not used.

An illustration of the explosion of the complexity of the topological structure when one passes from surfaces to 3-dimensional manifolds is given by the fact that there is only one prime surface (the torus), but there are infinitely many prime 3-dimensional manifolds. One still does not have a complete system of invariants to differentiate them.

At the end of his article [132], Milnor gave an example which shows that there is no analog of Theorem 19.2 in dimension 4. I will explain his example, because it uses two notions introduced in the sequel, the *blow ups of points* (see Chaps. 22 and 29) and the *stereographical projection* of a smooth quadric S of the complex projective space onto a complex projective plane P (see Chap. 29).

This stereographical projection is birational. It may be obtained by composing the blow up of the center of projection $O \in S$, which produces a complex projective surface M, and the contraction of the strict transforms E_1, E_2 on M of the two straight lines L_1, L_2 of S which pass through O. These two strict transforms are disjoint, as the blow-up of O separates L_1 and L_2. Denote by O_1 and O_2 the points of P obtained by contracting E_1 and E_2.

In M, one gets three smooth rational curves: E, E_1 and E_2. Each one of the three curves contracts to a smooth point of one of the surfaces S and P. Therefore, they have neighborhoods bounded by 3-dimensional spheres Σ, Σ_1, Σ_2: one may simply take preimages of balls centered at the smooth points O, O_1 and O_2. This allows one to decompose M in different ways as a connected sum, by cutting either along the sphere Σ or along the sphere Σ_1 and then by filling the resulting boundary components by 4-dimensional balls.

One may show that the closed 4-dimensional manifolds resulting from the neighborhoods of the complex rational curves E and E_1 are orientation-preserving diffeomorphic to the complex projective plane with the opposite orientation from the orientation induced by its complex structure, denoted $\overline{\mathbb{P}^2}$.

The two other closed 4-dimensional manifolds, obtained by filling with balls the complements of the neighborhoods of E and E_1, are diffeomorphic with the complex surfaces obtained by contracting E and E_1 respectively in M. That is, they are diffeomorphic with S and with the blow-up P' of P at O_2. The two previous contractions as well as the objects involved in the above explanation are represented schematically in Fig. 19.3.

Therefore, the two decompositions along the spheres Σ and Σ_1 allow one to write M in two different ways as a connected sum:

$$M = S \sharp \overline{\mathbb{P}^2} = P' \sharp \overline{\mathbb{P}^2}.$$

The point is that the 4-dimensional manifolds S and P' are not even homeomorphic, which shows that they cannot be decomposed into the same prime factors. In fact, in the language explained in Chap. 39, their intersection forms on their second homology groups $H_2(S, \mathbb{Z})$ and $H_2(P', \mathbb{Z})$ are not isomorphic. The first one takes only even values, which may be shown by using the fact that the smooth quadric S is doubly ruled, that is, algebraically isomorphic (therefore diffeomorphic) to $\mathbb{P}^1 \times \mathbb{P}^1$. This is not true for the second intersection form, which takes the value -1 on the homology class of the oriented surface E_2 in $H_2(P', \mathbb{Z})$.

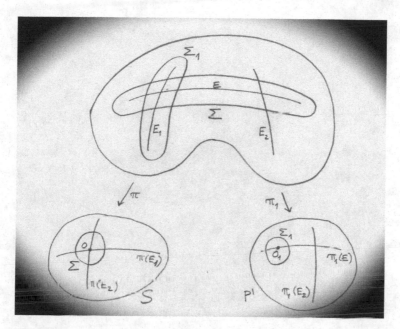

Fig. 19.3 Two different contractions

This shows that the two previous decompositions of M as connected sums cannot lead to decompositions into prime factors which coincide up to permutations.

Is there a notion of "ideal factors" in 4-dimensional topology, giving back the uniqueness of the decomposition?

After this brief incursion into higher dimensions, let us come back to algebraic curves and their associated Riemann surfaces.

Chapter 20
Clebsch and the Choice of the Term "Genus"

As we explained before, Riemann did not study algebraic curves embedded in the projective plane by themselves, but only up to birational transforms. Nevertheless, his techniques, passing through the associated Riemann surface or through the abelian integrals of the first kind attached to them, allow one to prove properties of such curves. For instance, one has:

Theorem 20.1 *A smooth algebraic curve of degree n in the complex projective plane has genus* $\dfrac{(n-1)(n-2)}{2}$.

In order to prove this theorem, one may use any of the following methods:

- The *Riemann–Hurwitz formula* (see Theorem 14.1). In our case, one projects the curve C onto a line L from a point of the projective plane situated in general position with respect to it. In this way, one obtains a covering of L of degree n, ramified at precisely $n(n-1)$ points, each one of them having ramification index 1. As L is topologically a sphere, it has genus 0, and the Riemann–Hurwitz formula applied to this ramified covering gives the desired formula for the genus of C.

- The *adjunction formula* (see Theorem 37.2), and the fact that the canonical class of the projective plane is -3 times the class of a straight line. This last fact may be seen by looking, for instance, at the form which is written $dx \wedge dy$ in an affine chart with coordinates (x, y). This form has a pole of order 3 along the line at infinity and no zeros.

The number p was called the "*genus*" by Clebsch in 1865, in the article [45], where he studied explicit questions about the geometry of plane projective curves[1]:

---

[1] I am grateful to Walter Neumann for the English translation of the fragments quoted in this chapter.

© Springer International Publishing Switzerland 2016
P. Popescu-Pampu, *What is the Genus?*, Lecture Notes in Mathematics 2162,
DOI 10.1007/978-3-319-42312-8_20

The class of Abelian functions to which an algebraic plane curve of degree n is related is determined by the number $p = \frac{(n-1)(n-2)}{2}$ if the curve has no double points [that is, points at which the curve may be defined by the equation $y^2 - x^2 = 0$ in certain local analytic coordinates] or cusps [that is, points at which the curve may be defined by the equation $y^2 - x^3 = 0$ in certain local analytic coordinates] [...]. If the curve has double points or cusps, then the value of p is reduced by the number of such points [...].

Instead of classifying the algebraic curves by their orders and, in every class obtained in this way, by the numbers of double points and of cusps which appear on them, one can subdivide them into Genera by the number p; in the first Genus all those for which $p = 0$; in the second Genus all those for which $p = 1$, etc. Then, conversely, the various orders appear as subdivisions in the Genera; and in fact every order in each genus up to $p = \frac{(n-1)(n-2)}{2}$, where find their place the most general curves of the n-th degree, that is, those which are free of double points and cusps.

Notice that in the previous definition, the genera are collections of curves, or classes, as is customary to say nowadays, not numbers. It is moreover a little strange that the n-th genus corresponds to $p = n + 1$. This shift was to be rapidly corrected by Clebsch, in his 1868 paper [46]:

I suggested to call the *genus* of a curve the number $p = \frac{1}{2}(n - 1)(n - 2) - d$ (Mr. Cayley's *deficiency*) where n is the order of the curve and d the number of its double points and cusps. The genera of two algebraic curves have to be the same in order to be able to establish a one-to-one correspondence between their points.

The phenomenon which appears here is that *the presence of singular points on a curve of degree n diminishes its genus (that is, the genus of the associated Riemann surface) compared to the one of a smooth curve of the same degree*. The simplest singular points are those which Clebsch called the "*double points*", where two smooth arcs intersect transversally. The singular point of the lemniscate is an example of a double point. In the sequel, we will rather use the terminology of "*ordinary double points*", because nowadays a "double point" means more generally any singular point of multiplicity 2.

By the way, what is the genus of the Riemann surface associated with the lemniscate? One has to search for its singular points in the whole complex projective plane, that is, *also its points on the line at infinity and its points with complex coordinates*. One discovers then that it has two supplementary ordinary double points at infinity.[2] Its genus is therefore $\frac{1}{2}(4 - 1)(4 - 2) - 3 = 0$. As explained by Clebsch (see Cayley's quotation in the next chapter), this shows that the lemniscate may be rationally parametrized!

It was Noether who explained, in his 1884 paper [143], the way in which arbitrarily complicated singularities diminish the genus of a curve of degree n (see Chap. 22). See also Chap. 33 for analogous questions in the world of algebraic surfaces.

---

[2]They are called the *cyclic points*, with homogeneous coordinates $[1 : \pm i : 0]$. All the circles of the euclidean plane with coordinates (x, y) pass through them, a fact which motivates their name.

Fig. 20.1 Acquiring an ordinary double point at the limit

To conclude this chapter, I would like to explain heuristically, using the topology of the associated Riemann surface, why an ordinary double point of an *irreducible* plane curve diminishes its genus by 1.

Let us imagine a sequence of smooth complex curves of degree n, which converge to a curve which has such a double point P. Then, near P, the curves of the sequence get ever narrower bottlenecks. At the limit, the circles which surround those necks shrink to the point P. One gets locally two disks having this point P in common (see Fig. 20.1; the two disks have spikes just on this 2-dimensional drawing, as in $\mathbb{C}^2$ they are smooth and they intersect transversally). When one constructs the associated Riemann surface, one abstractly pulls apart those disks: in the resulting surface there is a non-separating circle missing (the circle is non-separating because we assumed that the singular curve is irreducible). Therefore, the genus has dropped by one.

Chapter 21
Cayley and the Deficiency

One year after Clebsch's paper [45], Cayley published an article which cites it, but without retaining the term "genus" suggested by Clebsch Cayley introduced a rival term, the *"deficiency"*. This term was used for more or less half a century, mainly by British mathematicians, before being abandoned in favor of "genus". Here is the way in which Cayley explained the reason behind his choice (see [38, pp. 1–2]):

> The expression a "double point", or, as I shall for shortness call it, a "dp", is to be throughout understood to include a cusp [...].
>
> It was remarked by Cramer, in his "Théorie des Lignes Courbes" (1750), that a curve of the order n has at most $\frac{1}{2}(n-1)(n-2)$, $= \frac{1}{2}(n^2 - 3n) + 1$, dps.
>
> For several years past it has further been known that a curve such that the coordinates $(x : y : z)$ of any point thereof are as rational and integral functions of the order n of a variable parameter θ, is a curve of the order n having this maximum number $\frac{1}{2}(n-1)(n-2)$ of dps.
>
> The converse theorem is also true [...].
>
> The foregoing theorem, as a particular case of Riemann's general theorem, to be presently referred to, dates from the year 1857; but it was first explicitly stated only last year (1864) by Clebsch, in [45].
>
> The demonstration is, in fact, very simple; it depends merely on the remark that we may, through the $\frac{1}{2}(n-1)(n-2)$ dps, and through $2n-3$ other points on the given curve of the order n [...], draw a series of curves of the order $n-1$, given by an equation $U + \theta V = 0$, containing an arbitrary parameter θ; any such curve intersects the given curve in the dps, each counting as two points, in the $2n-3$ points, and in *one* other point; hence, as there is only one variable point of intersection, the coordinates of this point [...] are expressible rationally in terms of the parameter θ. [...]
>
> Before going further, it will be convenient to introduce the term "Deficiency", viz., a curve of the order n with $\frac{1}{2}(n-1)(n-2) - D$ dps, is said to have a deficiency $= D$: the foregoing theorem is that for curves with a deficiency $= 0$, the coordinates are expressible rationally in terms of a parameter θ. Since in such a curve the different points succeed each other in a certain definite order, viz, in the order obtained by giving to the parameter its different real values from $-\infty$ to $+\infty$, the curve may be termed a *unicursal* curve.

Let us detail a little more the proof, sketched by Cayley, of the fact that *any irreducible algebraic curve C of degree n contained in the projective plane and*

© Springer International Publishing Switzerland 2016

P. Popescu-Pampu, *What is the Genus?*, Lecture Notes in Mathematics 2162,
DOI 10.1007/978-3-319-42312-8_21

having exactly $\frac{1}{2}(n-1)(n-2)$ *ordinary double points as singularities, is necessarily rational.*

Consider a set E of $\frac{n(n+1)}{2} - 2$ points on C, consisting of its double points and of $2n - 3$ other points. Look now at the complex vector space H^{n-1} of homogeneous polynomials of degree $n - 1$ in three variables. It has dimension $\frac{n(n+1)}{2}$ and each point P of E defines a hyperplane of H^{n-1}, consisting of the polynomials which define a curve passing through P. As there are $\frac{n(n+1)}{2} - 2$ such hyperplanes, their intersection is of dimension at least 2. Therefore, one may find two *independent* polynomials U and V in this intersection. The curves defined by an equation of the form $U + \theta V = 0$ are therefore variable, and they all have degree $n - 1$. By Bézout's theorem,[1] their intersection number with C is always $n(n - 1)$. As C is assumed irreducible, the two curves never have common components. Consequently, this intersection number is the sum of the local intersection numbers at all the intersection points, and the result follows as explained by Cayley.

Note that the previous construction may be applied already for $n = 2$, that is, for a smooth conic section C. In this case E has only one element P. The curves defined by the equations $U + \theta V = 0$ are therefore the lines passing through P. One gets exactly the stereographical projection of C from P, already considered in Chap. 4 when we explained why smooth conic sections are rational curves.

By contrast with the situation concerning the complex plane projective curves, in higher dimensions one does not know how to express the maximal number of ordinary double points of a projective hypersurface as a function of its degree. Labs' thesis [125] contains details about the history of this problem.

---

[1]This theorem states that the number of intersection points, counted with multiplicities, of two complex plane projective curves without common components, is equal to the product of their degrees.

Chapter 22
Noether and the Adjoint Curves

As explained in the previous chapter, Cayley used the curves passing through the singular points of C in order to study a given plane projective curve C. Max Noether called such curves the *adjoints* of C. One has the following theorem, going back to Riemann [161], then refined by Clebsch, Gordan and Noether:

Theorem 22.1 *The genus of the Riemann surface associated with a plane projective curve C of degree n, with only ordinary double points as singularities, equals the dimension of the vector space of all polynomials defining adjoint curves of C of degree $n - 3$.*

Let us sketch a proof of this theorem. We work in an affine chart with coordinates (x, y) such that the corresponding line at infinity intersects C in n distinct points. Let T be the Riemann surface of C. Consider the restriction to T of a differential form of degree 1 in the plane, whose expression is:

$$q(x, y)\frac{dx}{\partial f/\partial y}$$

(expressions of this type are already fundamental in Abel's work [1]). One shows then by a computation in local coordinates that this restriction is everywhere *regular* (that is, it has no poles) on T if and only if the equation $q(x, y) = 0$ defines an adjoint curve of C of degree $n - 3$.

In his 1884 paper [143], Noether defined more generally the *adjoint curves* of a plane curve C having arbitrary singularities, as the curves C' with the following property: *for every point P of the curve C, and any affine chart (x, y) centered at P, one may define C' by an equation $q(x, y) = 0$ such that $q(x, y)\dfrac{dx}{\partial f/\partial y}$ is a regular differential form in restriction to the Riemann surface of C.* Then he showed that the previous theorem remains true in this more general context. He also analyzed the structure of the singular points using sequences of quadratic transformations and he proved the following generalization of Theorem 20.1:

© Springer International Publishing Switzerland 2016
P. Popescu-Pampu, *What is the Genus?*, Lecture Notes in Mathematics 2162,
DOI 10.1007/978-3-319-42312-8_22

Theorem 22.2 *The genus of the Riemann surface of an irreducible plane projective curve of degree n is equal to:*

$$\frac{1}{2}(n-1)(n-2) - M,$$

where M is the sum of all expressions $\frac{1}{2}m(m-1)$, when m varies among the multiplicities of all the singular points of the curve, counting the infinitely near ones.

Let us clarify the notion of an *infinitely near point* used in the previous statement. The points of the curve C situated in the ambient projective plane are considered to be its points in the usual sense. Let us consider one of them, denoted by A, and blow it up. The ambient projective plane becomes a new smooth projective surface, obtained by replacing A by a projective line L_A parametrizing the directions in the tangent plane at A. The points of L_A are said to be *infinitely near A*.

One may look at the curve $C \setminus A$ and take its closure in the new surface, getting the so-called *strict transform* C' of C. The points of C' situated on L_A are infinitely near points of C lying above A. For instance, if C has an ordinary double point at A, then there are two infinitely near points on $C' \cap L_A$, corresponding to the tangents of the two branches of C at A. Then one may repeat the same process at each point of $C' \cap L_A$, getting in an inductive fashion the whole set of infinitely near points of C which lie above A. This set is infinite, but it contains only a finite number of elements with multiplicity greater than 1. As a consequence, the sum M defined in Theorem 22.2 is necessarily finite.

More details about the previous theorem may be found in Brieskorn and Knörrer's book [25, Sect. 9.2, Theorem 5] and more details about the work of Max Noether can be found in the obituary [34] written by Castelnuovo, Enriques and Severi.

The previous theorem was extended in 1957 to abstract algebraic curves by Hironaka [94]. In order to arrive at that viewpoint, one first had to consider an *abstract* curve, that is, one which is independent of any ambient space. This was a gradual process, which began by viewing Riemann surfaces as abstract objects. The next chapter is dedicated to an explanation of this viewpoint.

Chapter 23
Klein, Weyl, and the Notion of an Abstract Surface

In 1882, Klein published his book [116] in which he tried to explain the physical intuitions lying behind Riemann's theory. In the introduction, he argued that Riemann's surfaces do not necessarily arise from functions, but that one may reverse this process:

> I am not sure that I should ever have reached a well-defined conception of the whole subject, had not Herr Prym, many years ago (1874), in the course of an opportune conversation, made me a communication which has increased in importance to me the longer I have thought over the matter. He told me that *Riemann's surfaces originally are not necessarily many-sheeted surfaces over the plane, but that, on the contrary, complex functions of position can be studied on arbitrarily given curved surfaces in exactly the same way as on the surfaces over the plane.*

If one reads Klein's book, one understands that his *"arbitrarily given curved surfaces"* are in fact surfaces in 3-dimensional space. The vision developed more with Weyl, who gave for the first time in his 1912 book [191, I, Sect. 4–6] an *abstract* definition of a surface, without any reference to an ambient space. The explanation of his motivations, given in the preface of the book, is a variation of Klein's, as Prym is simply replaced by Klein:

> Klein had been the first to develop the freer conception of a Riemann surface, in which the surface is no longer a covering of the complex plane; thereby he endowed Riemann's basic ideas with their full power. It was my fortune to discuss this thoroughly with Klein in diverse conversations. I shared his conviction that Riemann surfaces are not merely a device for visualizing the many-valuedness of analytic functions, but rather an indispensable essential component of the theory; not a supplement, more or less artificially distilled from the functions, but their native land, the only soil in which the functions grow and thrive.

Weyl defined a surface as a *separated topological space locally homeomorphic to* $\mathbb{R}^2$. At that time, these concepts were just emerging. For this reason, he also

P. Popescu-Pampu, *What is the Genus?*, Lecture Notes in Mathematics 2162,
DOI 10.1007/978-3-319-42312-8_23

explained basic notions of general topology.[1] He emphasized the advantages of the
chosen axioms, emerging from an analysis of the ideas which are involved in the
process of analytic continuation. As we saw in Chap. 12, that process forces us to
think about small disks which progressively cover parts of the plane.

Notice also how, if one passes from Klein's title to Weyl's, attention moves from
algebraic functions and their integrals to Riemann surfaces, in agreement with the
philosophy presented in the previous quotation. Later, essentially starting from the
end of the 1920s, a new vision of abstract varieties endowed with various types of
structures was developed: *topological*, *differentiable* with various differentiability
classes, *real analytic*, *complex analytic* (or *holomorphic*), *Riemannian*, *Kähler*,
symplectic, etc.

---

[1]The language and the definitions were not necessarily the same as ours. The first axiomatic
presentation of a large class of topological spaces was to appear shortly, in Hausdorff's 1914 book
[91]. One may consult Moore's paper [137] in order to see the complicated evolution of the basic
concepts of general topology.

Chapter 24
The Uniformization of Riemann Surfaces

Riemann surfaces were introduced to build new domains of definition on which multivalued algebraic functions become univalued. One of the advantages of algebraic functions which was lost in this procedure was the ability to use *a single parameter* in order to describe the function in its full domain of definition (which becomes the whole Riemann surface). In fact, this advantage is still available if one lifts the function to the *universal covering* of the Riemann surface, as shown by the following *uniformization* theorem:

Theorem 24.1 *Any abstract simply connected Riemann surface is isomorphic to the Riemann sphere, to the complex plane, or to the unit disk.*

Let us explain briefly the intimately related notions of *simple connectedness* and *universal covering*. In Chap. 12 we saw that it was possible to measure the multivaluedness of a function by looking at the way its various determinations are permuted when one travels along a loop. Moreover, the resulting permutation depends only on the homotopy class of the loop, provided its base point remains fixed during the homotopy. This fact, first formulated by Cauchy and Puiseux for algebraic functions of one complex variable, is general. It led Poincaré to introduce the class of connected topological spaces in which any loop is homotopic to a constant one (see Chap. 38). One says that such spaces are *simply connected*.

We saw that the complement in the Riemann sphere $\overline{\mathbb{C}}$ of the set of ramification points of an algebraic function is covered by several sheets of the Riemann surface associated with the function. Similarly, any topological space X which is simple enough both locally and globally (it is sufficient if it is a paracompact topological manifold) admits a simply connected covering. One may prove that such a covering is unique (technically speaking, only up to an isomorphism of coverings). One calls it *the universal covering* of X.

Let us come back to the case of a Riemann surface T. If one uses the same parameter t everywhere on the universal covering, which may be assumed to be included in $\overline{\mathbb{C}}$ according to the previous theorem, one may regard any meromorphic

© Springer International Publishing Switzerland 2016
P. Popescu-Pampu, *What is the Genus?*, Lecture Notes in Mathematics 2162,
DOI 10.1007/978-3-319-42312-8_24

function on T as a *uniform* function of the parameter t. This is the reason why one speaks about a *uniformization theorem*.

This theorem, which was conjectured by Poincaré around 1880 and finally proved completely rigorously around 1908 by Poincaré on one side and by Koebe on another, has a complex and profuse history, which is related in the book [52] by de Saint-Gervais.

A trichotomy among the smooth algebraic curves appears in this way:

- those of genus 0 are their own universal covers, isomorphic to the Riemann sphere;
- those of genus 1 are universally covered by the complex plane;
- those of genus at least 2 are universally covered by the unit disk.

This trichotomy is also very important in number theory (see Theorem 27.1).

In the next chapter we focus on some number theoretical aspects of the genus of curves.

Chapter 25
The Genus and the Arithmetic of Curves

In the introduction of his 1928 thesis [183], Weil explained the usefulness of the notion of genus in arithmetic:

> The object of geometry on an algebraic curve is the study of those properties of points and systems of points on the curve which are invariant under birational transformations. [...] In particular, the search for rational points on a given curve C is obviously a problem which is invariant under birational transformations with rational coefficients, and belongs therefore to the arithmetic on algebraic curves: when the domain of rationality is reduced to the set of rational numbers, this problem is nothing else than that of resolution in rational numbers of Diophantine equations in two variables or (which is the same), that of resolution in integers of *homogeneous* Diophantine equations in three variables.
>
> Since Diophantus, who left them his name, plenty of special types of equations of this kind have been studied, and some of them have aroused considerable efforts: it will be enough to cite the equation $x^n + y^n = 1$, whose impossibility in rational numbers for $n > 2$, stated by Fermat in his "Observations sur Diophante", has remained unproved till now.[1] But it is only at a very recent epoch that the progress of geometry on algebraic curves suggested to treat by analogous methods the general study of Diophantine equations in two variables. Hilbert and Hurwitz were the first to notice that the search for rational points on an algebraic curve is a problem which is invariant under birational transformations with rational coefficients: as a consequence, the fundamental element of classification for Diophantine equations in two variables is the genus of the equation, and not its degree.

His *"homogeneous Diophantine equations in three variables"* are those of the form $P = 0$, where the polynomial $P \in \mathbb{Z}[X, Y, Z]$ is homogeneous. Weil explained in the following way the problem which he solved in his thesis, emphasizing the key role played by systems of p points on the curves of genus p defined by such equations:

> [...] the greatest part of this essay [[150], by Poincaré] is devoted to the study of rational points on the curves of genus 1, and especially on the cubics. Its most important contribution is the definition of the *rank* of a curve of genus 1 with rational coefficients [...] one may

---

[1] It was to be proved only in the 1990s by Wiles [194].

© Springer International Publishing Switzerland 2016

P. Popescu-Pampu, *What is the Genus?*, Lecture Notes in Mathematics 2162,
DOI 10.1007/978-3-319-42312-8_25

define it, briefly, as the minimal number of rational points on the curve from which all the other ones may be deduced by rational operations.

In the last paragraph of his essay, in which he considers the curves of arbitrary genus p, Poincaré shows that, in order to generalize the results obtained for genus 1, one has to consider not the rational points on the curve, but instead the rational systems of p points: using them, he defines an invariant under the birational transforms with rational coefficients, the *rank*, which is the minimal number of rational systems of p points from which all the other ones may be deduced through rational operations. Since Poincaré, the most important progress has been achieved by Mordell, who proved that the rank of the curves of genus 1 is necessarily finite when the domain of rationality reduces to the set of rational numbers. [...]

In the present work, I prove that the rank of a curve C is finite for any genus p and for any number field (algebraic and finite) which is chosen as domain of rationality.

The desire to relate algebraic geometry and arithmetic was to be a constant of the mathematical work of Weil (see also his lecture [185] from 1950). We will come back to this topic in Chap. 43.

Chapter 26
Several Historical Considerations by Weil

In order to have an overview of the path we have traveled so far, let us discover the viewpoint described by Weil [190] in 1981:

> Briefly and roughly speaking, it is allowed to say that algebraic geometry is the study of equations and systems of algebraic equations in several variables when the sets determined in this way (also called "varieties") are not reduced to isolated points. Typically, the problem is formulated relatively to a base field which may or may not be specified; when this field is chosen according to its arithmetic properties, or even more so when one works over such or such a ring instead of over a field, one comes so close to the rather hazy frontier between number theory and algebraic geometry that one barely notices to have crossed it; by the way, it is of no importance to know whether one indeed crossed it.
>
> Defined in this way, algebraic geometry is inseparable from the theory of algebraic functions of one or of several variables which, since the nineteenth century, has constituted an important chapter of it; it is convenient to append to it also the theory of integrals of algebraic differentials (at least if one is working over the field of complex numbers, as was customary throughout the nineteenth century and beyond), or at least the theory of those differentials which keep a meaning for every base field.
>
> Of course, a mathematical discipline is no less characterized by its methods than by its object. In this respect, what matters above all for the practice of algebraic geometry as we understand it, is the use of rational transformations and especially of birational transformations, which contain as special cases the affine and projective transformations. Secondly, it is the classification of the studied objects with respect to the considered transformations, for instance (when one deals with algebraic curves) first using the degree, then using the genus. Sometimes, one may avoid the geometrical language by the direct study of the fields of algebraic functions but, in the absence of this language and of the links or at least of the analogies which he puts into evidence with topology and differential geometry, it is to be believed that this subject would have dried long before.

After having presented the example of integration given by Johann Bernoulli, as shown in our Chap. 4, Weil gave several examples of the appearance of this type of problem:

> We find later the same idea taken up by Daniel, son of Johann Bernoulli, in a letter from 1723 to Goldbach [...]:
>
> [...] "Those Diophantine problems are often of a great use in the integration of differential expressions, and I employed them often in problems of integration, a reason

© Springer International Publishing Switzerland 2016
P. Popescu-Pampu, *What is the Genus?*, Lecture Notes in Mathematics 2162,
DOI 10.1007/978-3-319-42312-8_26

why I am surprised that they were so little developed. In the past they were discussed by
Roberval, Wallis, Fermat, etc. and even with an excessive enthusiasm, in spite of the fact
that they did not know their use."

[...] Likewise also d'Alembert, who expresses himself on this subject with his usual
lucidity in [...] the *Encyclopédie*:

Let us note by the way that this method to reduce to rational quantities the irrational ones
is very useful in the integral calculus in order to reduce a given differential to a rational
fraction.

It is visibly equations of genus 0 that d'Alembert and the Bernoullis have in mind in the
fragments just quoted. Leibniz, as always, was seeing those topics from a higher viewpoint
when he was writing in 1702 [...]:

[...] there will appear [mathematicians], I hope so, who will spread at a larger scale the
seeds of the new theory and who will harvest more abundant fruits, which will be especially
the case if they endeavour, more than it was the case till now, to develop the Diophantine
Algebra, almost completely neglected by the disciples of Descartes because they did not
see its use in geometry. On the contrary, I remember having indicated several times (which
could seem surprising) that a good portion of the progress of our integral calculus depends
on the extension of the sort of Arithmetic that Diophantus treated systematically, as far as I
know, for the first time.

As may be seen from these prophetical lines, not only had Leibniz perceived the integrals
of algebraic differentials (precisely those which were lately called abelian) as a privileged
field for the future development of integral calculus, but he had also seen clearly the narrow
relation between this subject and the classical "Diophantine algebra", that is, in short,
algebraic geometry. As for the questions about the classification of integrals, Leibniz was
interested in them early on, as may be seen in his letter from 1677 to Oldenburg [...]:

[...] "Is still missing a way to recognize whether the quadrature of a given figure may
be brought to that of the circle or of the hyperbola. Because the majority of the figures
which were treated till now could be squared in this way. But if, as I believe it, one may
show that this is not always possible, then remain to be found figures of a higher genus, to
which may be reduced the quadratures of all the other ones... Gregory was thinking that
the rectification of the hyperbola and of the ellipse did not depend on the quadrature of the
circle or of the hyperbola..."

The quadratures of the circle and the hyperbola are the inverse circular functions and
the logarithmic ones; the quadratures which depend on them are the integrals of genus 0.
Gregory's conjecture was justified; in the rectification of the ellipse and of the hyperbola,
one meets an "elliptic" integral of the form

$$\int \sqrt{\frac{1 - k^2 x^2}{1 - x^2}}\, dx;$$

the geometers of the seventeenth century soon added to them the arc of the lemniscate

$$\int \frac{dx}{\sqrt{1 - x^4}}.$$

Leibniz, followed by Joh. Bernoulli, never completely lost the hope to bring those
integrals back to the inverse circular and logarithmic functions [...]. Even Euler had not
given up in 1730; at least he asks twice the question in letters to Goldbach [...], and one
may see here a first indication of the passionate interest he was to show in the sequel for
elliptic integrals. [...]

For the sake of the reader with a strong historical curiosity, I would like to indicate two other historical references. Brill and Noether described in a very detailed way in their 1894 paper [180] their vision of the historical development of the theory of algebraic functions and curves. Abhyankar offered in his 1876 paper [4] a pleasant historical ramble through the world of algebraic plane curves, insisting on the usefulness of keeping an "elementary" viewpoint.

Chapter 27
And More Recently?

Summarizing what we have seen so far, the *genus* is a measure of the complexity of abstract objects: algebraic curves (defined over an arbitrary field, etc.) and topological surfaces (orientable, etc.) This is in contrast with the *degree*, which partially measures the embedding into an ambient space.

To conclude the part of this story which deals with algebraic curves and topological surfaces, I would like to state two more recent very deep theorems concerning the genus, seen from both viewpoints. The first one is the affirmative answer given by Faltings [75] to a conjecture of Mordell, and the second one is the affirmative answer given by Kronheimer and Mrowka [124] to a conjecture of Thom.

Theorem 27.1 *Any irreducible curve of genus $p > 1$ which is defined by equations with rational coefficients admits a* finite *number of points with rational coordinates.*

This is to be contrasted with the curves of genera 0 or 1, which may have an infinite number of points with rational coordinates (see Chap. 25).

Theorem 27.2 *If a smooth, compact, connected and oriented surface without boundary is embedded in the complex projective plane and if it is homologous there to a smooth complex algebraic curve of degree n, then its genus is at least equal to that of the curve, that is, to $\frac{(n-1)(n-2)}{2}$.*

Recall that we stated the previous formula for the genus of a smooth algebraic curve of degree n in Theorem 20.1. In what concerns the notion of *homology*, it is treated briefly in Chaps. 38 and 39. Note that all the smooth algebraic curves of the same degree are not only homologous, but also continuously deformable into one another inside the complex projective plane, that is, they are *isotopic*.

The previous theorem is perhaps evidence of a certain *"principle of economy of algebraic geometry"*, still rather mysterious, which states that in order to construct some kinds of objects, one cannot proceed in a simpler way (here, with a smaller genus) than using algebraic geometry.

© Springer International Publishing Switzerland 2016
P. Popescu-Pampu, *What is the Genus?*, Lecture Notes in Mathematics 2162,
DOI 10.1007/978-3-319-42312-8_27

Part II
Algebraic Surfaces

Chapter 28
The Beginnings of a Theory of Algebraic Surfaces

The works of Riemann had a great impact on his contemporaries, so much that during the rest of the nineteenth century, with few exceptions,[1] algebraic geometers directed their efforts toward the development of *complex* algebraic geometry, that is, the study of the geometry of the sets of complex solutions to systems of polynomial equations in several variables.

In this context, *algebraic surface* meant *complex algebraic surface*, that is, a surface which may be locally parametrized by functions in two variables, in the neighborhood of any non-singular point. Analogously, as Riemann surfaces need only one complex parameter in order to be described locally, they are complex *curves* rather than complex *surfaces*!

It was mainly Picard who, at the end of the nineteenth century, developed the transcendental viewpoint of studying integrals of algebraic forms of degrees 1 and 2 on algebraic surfaces. His approach was rooted in a *tomographical* technique, consisting in slicing the surfaces of affine space by parallel planes, which allowed him to regard them as families of algebraic curves depending rationally on a complex parameter, and to use Riemann's theory on the associated Riemann surfaces. Nowadays one speaks of the "Picard–Lefschetz theory" and of the "Picard–Fuchs equations" for various aspects of the ideas which developed from this method. The interested reader may consult Picard's treatise [147], as well as the introduction [105, Chap. X] to his works.

Several of the theorems about Riemann surfaces and algebraic curves formulated in the first part of this book deal with the *meromorphic* functions on them, and the associated loci of zeros and poles. No order relation on those sets of points entered into play. It thus became practical to denote them additively, e.g. $\sum_i A_i$, $\sum_j B_j$. Such sums were called *groups of points*.

---

[1]Such an exception in described in Chap. 48.

© Springer International Publishing Switzerland 2016
P. Popescu-Pampu, *What is the Genus?*, Lecture Notes in Mathematics 2162,
DOI 10.1007/978-3-319-42312-8_28

Nowadays, one says that both are *effective divisors*. If such a divisor contains
m points, one says that it has *degree m*. That is, an effective divisor of degree m
is a formal sum of m (possibly coinciding) points of the Riemann surface. If the
groups of points $\sum_i A_i$, $\sum_j B_j$ were related as before (that is, one is the set of zeros
and the other the set of poles of the same meromorphic function, both counted with
multiplicities), then they were called *linearly equivalent*.

Theorems 17.1 and 17.3 of Abel show that the study of sums of abelian
integrals (perceived as *transcendental* objects, depending on the integral calculus)
is intimately related to that of groups of points up to linear equivalence (perceived
as *geometrical* objects).

Brill and Noether [179] rebuilt Riemann's theory on geometric bases, banishing
the use of integration, and taking as the central objects of their study of algebraic
curves the *linear series*, that is, projective spaces of linearly equivalent groups of
points. Such linear series are called *complete* if they contain all the effective divisors
which are linearly equivalent to one of their divisors. The aim of their theory was
to reprove as many of Riemann's results as possible by geometric means and, of
course, to discover new ones.

Similarly, the geometric viewpoint on the theory of algebraic surfaces consisted
in studying them using *linear systems* of hypersurfaces of the ambient projective
space, that is, hypersurfaces defined by equations of the form:

$$\lambda_1 F_1 + \cdots + \lambda_m F_m = 0$$

where $F_1, \ldots, F_m$ denote homogeneous polynomials of the same degree and where
the parameters $\lambda_1, \ldots, \lambda_m$ do not vanish simultaneously. These parameters are
therefore present *linearly* in the equation, which explains the name of this type of
family. The intersections of the members of such linear systems of hypersurfaces
with the initial surface formed *linear systems* of curves on the surface, which were
analogs of the linear series of groups of points on algebraic curves.

One has a notion of a *complete linear system*, whose definition mimics that of a
complete linear series of points on a curve. If C is a curve on a surface, the complete
linear system containing it is denoted by $|C|$.

Let us come to the problem of defining a notion of genus for algebraic surfaces.
Given its importance in the works of Riemann, it is natural to expect that an
analogous notion also plays a key role in the theory of complex algebraic surfaces.

One of the essential properties of the genus of a plane curve is its birational
invariance (see Chap. 15). From the beginning, this property was taken as a guide
for the construction of an analogous notion of genus for algebraic surfaces contained
in 3-dimensional projective space. Clebsch [46] gave such a definition in 1868, by
imitating the definition of the genus of a plane curve through *adjoint curves* (see
Chap. 22):

I managed to prove the following theorem:
"Let $f = 0$ and $\phi = 0$ define two surfaces of orders m and n respectively. Assume that
one may associate to each point of $f = 0$ a single point of $\phi = 0$ and conversely, in such
a way that those surfaces are transformable one into the other in an algebraic and rational

way. Assume, for more simplicity, that those surfaces have only regular singularities [...].
Let then p denote the number of arbitrary coefficients of a surface of order $n - 4$ which
passes through the double curves and the cuspidal curves lying on $f = 0$, and let p' be the
corresponding number relative to $\phi = 0$. One will always have $p' = p$."

This theorem allows us to classify those surfaces with respect to their genus p.

Slightly later, and mainly due to Noether's works [141, 142], a new notion of
genus for surfaces was developed. Castelnuovo and Enriques explained it in the
following way in 1897, in their survey paper [32, Sect. 3–4]:

With respect to the birational transformations, the algebraic varieties with a given number of
dimensions get distributed into various classes; we will say, as Riemann, that two varieties
belong to the same class when it is possible to establish a birational correspondence between
them. [...]

It is well-known how Riemann, studying from the viewpoint of *Analysis situs* the real
surface which represents the points (real and imaginary) of an *algebraic curve*, attained the
most important of those characters, which he called the *genus of the curve*[2] [...] Clebsch
and Gordan, when they explained the same conception under a geometric form, gave the
following well-known definition of the genus. [...][3]

Alongside this *geometrical* definition one may give a *numerical* definition of p. It is
enough to notice that the number of parameters which enter homogeneously in the equation
of a curve of order $n - 3$, compelled to pass through d given points, is:

$$p = \frac{(n-1)(n-2)}{2} - d.$$

Here, in fact we assume that the d double points of C impose as many *distinct* conditions on
the curves of order $n - 3$ compelled to pass through them; this is not a priori obvious, but it
was proved later (Brill and Nöther) that this assumption is always satisfied, when restricted
to irreducible curves.

The extension of those definitions to algebraic surfaces is immediate, in the simplest
cases; Clebsch and Mr. Nöther indicated it.

Let F be an algebraic surface of order n in ordinary space; assume that F does not have
other singularities than a double curve of order d (≥ 0) and genus π, and a certain finite
number t (≥ 0) of triple points of the surface which are also triple points of the double curve
considered before. Next, consider the surfaces of order $n - 4$ which pass simply through the
double curve (and therefore twice, in general, through its triple points), surfaces which we
will call *adjoints* to F; *the number of such surfaces which are linearly distinct* is (according
to the proof of Mr. Nöther) an invariant character (relative to the birational transformations)
of the surface F; one calls it the *geometric genus* of F, and one denotes it by p_g.

Now, if we notice that a surface of order m *big enough* must satisfy

$$md - 2t - \pi + 1$$

conditions in order to contain the double curve of F, we are led (with Cayley), to form the
expression

$$p_n = \frac{(n-1)(n-2)(n-3)}{6} - (n-4)d + 2t + \pi - 1,$$

---

[2]This is a historical imprecision. Indeed, we saw in Chap. 20 that it was Clebsch who introduced
this name.

[3]Here Castelnuovo and Enriques explain Theorem 22.1.

whose value is exactly p_g, if $m = n-4$ is so big that one may apply the previous formula. Be the last condition verified or not, the *expression for p_n always has the invariance property*; this was shown by Zeuthen[4] and Nöther. This expression is perfectly analogous to the one considered above when speaking about plane curves; the name of *numerical genus* of the surface F was given to it.

But the analogy with plane curves stops here; indeed, we can no longer assert the equality between the values of p_g and p_n; instead, *it is possible that $p_g \neq p_n$ and more precisely $p_g > p_n$*. For instance, a ruled surface whose plane sections have genus p has (following Cayley),[5] $p_g = 0$, $p_n = -p$. [...]

A little later, it became customary to call *the arithmetic genus* what Castelnuovo and Enriques had called *the numerical genus*. This gave rise to a dichotomy among complex surfaces:

- some of them, called *regular*, satisfy $p_g = p_n$;
- the other ones, called *irregular*, satisfy $p_g > p_n$.

The efforts of Humbert, Enriques, Castelnuovo, Severi, Picard and Poincaré (see [195, Chap. VII.3]) at the beginning of the twentieth century were to lead to the proof that the *irregularity $p_g - p_n$* of an algebraic surface may also be seen as a generalization of the notion of the genus of an algebraic curve (interpreted via Theorem 17.2).

Namely, they showed that *the irregularity is equal to the dimension of the complex vector space of the holomorphic differential forms of degree 1 which live on the algebraic surface.*

---

[4]In the article [200].

[5]The precise reference is [39].

Chapter 29
The Problem of the Singular Locus

We saw in the last chapter's quotations that the first definitions of notions of genus for algebraic surfaces were formulated for those contained in the three-dimensional complex projective space $\mathbb{P}^3$. One might think that this was caused by a reluctance to consider surfaces in a higher-dimensional space, where geometric intuition was undermined.

However, it is possible to argue that there is another reason, namely, as explained from the start by Castelnuovo and Enriques, that the aim of the researchers was to develop a *birationally invariant* theory of surfaces, in line with the curve theory elaborated by Riemann (see Chap. 15). It turns out that any algebraic surface from a projective space of arbitrary dimension is birationally equivalent to a surface in $\mathbb{P}^3$: it is enough to take a generic linear projection.

The problem is that such representatives inside $\mathbb{P}^3$ of a birational equivalence class may have *singularities*, even if the surface which is projected generically is smooth. This explains the reason why the singular locus was always taken into account in the definitions. Moreover, Castelnuovo and Enriques [32, page 275] see in it a tool in the search for *invariants*:

> One arrived in the first place at the notion of invariants by considering the transcendental expressions associated with the surface (Clebsch, Nöther 1868–69; Picard 1884). Then one intended to define the same invariants geometrically (or algebraically). The geometers who approached this question (Cayley, Nöther, Zeuthen) proceeded in the following way. They considered a surface F of ordinary space, and they payed attention to its projective characters (order, multiple curves and points). They then transformed F birationally into another surface F' of the same space, and by comparing the projective characters of F with the analogous characters of F', they formed numerical expressions or functions which were not modified by the transformation.

This is analogous to the way knot invariants are constructed from their diagrams, which also contain singularities (their crossing points), amply used in the twentieth century.

© Springer International Publishing Switzerland 2016
P. Popescu-Pampu, *What is the Genus?*, Lecture Notes in Mathematics 2162,
DOI 10.1007/978-3-319-42312-8_29

However, once a construction which used the singular locus of a surface was reached, it was important to understand how it could be expressed on a smooth representative of its birational equivalence class. Here is what Castelnuovo and Enriques wrote in [32, Chap. 3] regarding this subject:

> Relative to the birational transforms, the algebraic varieties of the same dimension get distributed into classes [...].
>
> It is true that the varieties of a same class differ by their projective characters (order, dimension of the space to which they belong, ...); but they have various common properties, whose research constitutes the subject of the *Geometry of algebraic varieties* [...]. It follows that, when one studies a *class* of varieties, one may always make reference to an abstract general variety of the class, without paying attention at all to its projective characters [...].
>
> The consideration of the general variety of a class has this great advantage, that it allows to make abstraction of all the singularities of projective nature which a particular variety may present. Thus, in the sequel, when we speak about the points of the general variety of a class, we will always assume that they are *simple*; of course, we do not exclude that such points may correspond to multiple points on particular varieties of the class; but we simply state that a multiple point is a projective feature of some representative of the class, not of the class itself; it is, so to speak, a defect of the image, not of the model.

In particular, the authors state that *any projective variety is birationally equivalent to a smooth projective variety*. It was known how to prove this for projective curves, and various approaches appeared from time to time for projective surfaces, each one of them subsequently arousing criticism. One may consult Zariski's book [195, Chap. I] for a description of the main works concerning this question up to 1935, as well as Gario's papers [82] and [83] for details about the discussions around this problem in Italy in the 1890s.

In 1935, in his monograph [195] presenting the theory of algebraic surfaces from the Italian viewpoint, Zariski considered that the first completely rigorous proof of the fact that *every projective surface is birationally equivalent to a smooth one* had just been obtained by Walker in [181]. From this moment on, Zariski tried to extend this theorem to higher dimensions. He could do it in [198] for the projective varieties of dimension 3, but it was only his student Hironaka who finally proved in his 1964 paper [95] the analogous theorem for any algebraic variety defined on a field of characteristic zero[1] (see Theorem 29.1).

Let us come back to the survey paper of Castelnuovo and Enriques. Here is how they described in [32, Chap. 2] the difficulties which appear in birational geometry of dimension at least two, even if one works only with non-singular varieties:

> Coming back to the varieties F and F' in birational correspondence, there are two cases to be considered, according to the situation in which F, F' are either curves, or instead varieties of more than one dimension. In the first case indeed, there is a biunivocal correspondence *without exceptions* between the curves F, F' [...]. But if F, F' are for instance surfaces [...], there may exist simple points (in finite number) on F or F', to each one of them corresponding on F' (or F) all the points of a curve. The stereographical projection of a surface of the second order on a plane gives a simple example of such curves.

---

[1]This remains an open question today in positive characteristics, for varieties of dimension at least 4.

We explain now this last example, already used in Chap. 19. The *stereographical projection* is usually defined as the projection of a sphere of the standard euclidean 3-dimensional space, from one of its points interpreted as a pole, to the tangent plane at the diametrically opposite pole.

But we have to think here about this operation both projectively and over the field of complex numbers. The sphere is then indistinguishable from the other non-singular quadrics, whose real affine models are the *ellipsoids*, the *hyperboloids* of one or two sheets and the elliptic or hyperbolic *paraboloids*. Indeed, up to a projective transformation, there is a single smooth quadric in the complex projective space, a result which is valid in all dimensions.

To project stereographically such a quadric means to project it from one of its points to a hyperplane which does not contain it. We have already seen in Chap. 4 that, using this type of projection, one may show that a conic is birationally equivalent to a line. In fact, in all dimensions, *the stereographical projection establishes a birational equivalence between a smooth quadric and the hyperplane of projection.*

Let us explain this fact with the example of a hyperbolic paraboloid S. We chose this real affine model in order to understand better what happens, with the help of our intuition of the usual 3-dimensional real affine space, and at the same time in order to follow easily the operations using computations.

One may choose an affine system of coordinates such that S is defined in the space $\mathbb{R}^3$ by the equation $z = xy$ and such that the center of projection is the origin O. As a plane of projection, let us choose the one defined by the equation $x + y = 1$ (see Fig. 29.1). Denote it by P. If $A_0(x_0, y_0, z_0) \in S$, then the line which joins O and A_0 intersects the plane P at the point $B_0(\frac{x_0}{x_0+y_0}, \frac{y_0}{x_0+y_0}, \frac{z_0}{x_0+y_0})$.

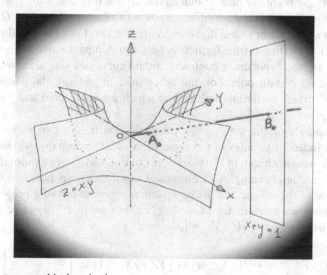

Fig. 29.1 A stereographical projection

By using the coordinate system (x, y) on S and the coordinate system (x, z) on P, which we will denote by (X, Z) in order to avoid confusion, our stereographical projection σ is given by:

$$(x, y) \longrightarrow (X, Z) = \left(\frac{x}{x+y}, \frac{xy}{x+y} \right).$$

One deduces straight away the expression for the inverse σ^{-1}:

$$(X, Z) \longrightarrow (x, y) = \left(\frac{Z}{1-X}, \frac{Z}{X} \right),$$

which indeed shows that σ is birational. Geometrically, this comes from the fact that, because the quadric S has degree two, *every* line passing through the center of projection O cuts it at a single other point, possibly coinciding with O or lying at infinity.

Really, *every* line? Well, almost every, the only exceptions being the lines passing through O and *contained in S*.

In fact, through every point of S we can find two such lines: one says that the quadric is *doubly ruled*. This phenomenon may also be understood geometrically: the tangent plane at a point A of S intersects S along a conic which is singular at A, and is therefore the union of two lines. Figure 29.1 shows those which pass through the origin, which are exactly the x and y coordinate axes.

Therefore, the projection σ maps the two lines of S passing through O to two distinct points M and N of P. Besides, σ is not defined at O, because one obtains as possible limits of the points $\sigma(A)$, when A tends towards O, all the points of the line L which joins the points M and N: indeed the inverse σ^{-1} sends the line L, defined by the equation $Z = 0$ in the coordinate system (X, Z), into the origin O. One says that σ *blows up* the point O and that it *contracts* the two lines passing through O.

Whenever a birational transformation between compact smooth algebraic surfaces is not an isomorphism, it contracts certain curves on one side, which means that it blows up certain points on the other side. In general, the phenomenon is mixed, each surface including both curves which get contracted and points which get blown up.

In the example of the stereographical projection σ, the two surfaces are not isomorphic. Indeed, two lines of the same ruling of a smooth quadric of $\mathbb{P}^3$ do not intersect; on the other hand, in $\mathbb{P}^2$ two distinct lines always have a point in common.

But there are also examples of birational transformations between isomorphic surfaces. One of the simplest is the *quadratic transformation* of a projective plane with homogeneous coordinates $[X : Y : Z]$ into another projective plane with homogeneous coordinates $[U : V : W]$, defined by:

$$[X : Y : Z] \dashrightarrow [U : V : W] = [YZ : ZX : XY] = \left[\frac{1}{X} : \frac{1}{Y} : \frac{1}{Z} \right].$$

This transformation "contracts" the sides of the triangle defined by the equation $XYZ = 0$ into the vertices of the triangle defined by $UVW = 0$ and at the same time it "blows up" the vertices of the first triangle into the sides of the second one.

The birational transformations were thus always seen as more or less intricate combinations of blow ups and contractions. However, in his 1942 paper [197], Zariski successfully defined a notion of *blow up of a point*, allowing any birational transformation between smooth surfaces to be factored as a product of such blow ups and their inverses.

More generally, and in all dimensions, he defined the notion of the *blow up of a smooth subvariety M* of a smooth projective variety N: in modern terms, one replaces every point of the subvariety M by the projectified normal space to M in N at that point (that is, by the projectivization of the quotient of the tangent space to N by the tangent space to M).

Zariski also showed that this definition could be used to get a notion of the *blow up* of a smooth subvariety M of a not necessarily smooth projective variety N: one has just to embed N in a smooth ambient space and to blow M up there. It may be shown that, in restriction to N, this gives a construction which is independent of the chosen embedding.

Zariski's isolation of a general blow up operation led him to contemplate the possibility of an algorithm for the resolution of singularities *by successive blow ups of smooth subvarieties*. It was Hironaka who finally proved such a result in his 1964 paper [95]:

Theorem 29.1 *Let X be an algebraic variety defined over an algebraically closed field of characteristic zero. Then there exists a resolution of singularities of X, that is, a proper birational morphism $\rho: Y \rightarrow X$, which is an isomorphism over the smooth part of X. Moreover, such a morphism ρ may be obtained by composing blow ups of smooth varieties.*

Since then, this very deep theorem has become a fundamental tool in the study of singular algebraic varieties.

Chapter 30
A Profusion of Genera for Surfaces

The Italian school of algebraic geometry of surfaces, whose most famous representatives were Castelnuovo, Enriques and Severi, privileged geometric methods when studying a surface, via the algebraic families of curves it contained. The simplest such families are those which depend linearly on the parameters: the *linear systems of curves*, which generalized the linear series of groups of points on algebraic curves, considered by Brill and Noether (see the beginning of Chap. 28).

The relationship between this viewpoint and the transcendental one which starts from simple or double integrals is supplied by the *zero-loci* of holomorphic two-forms. These give rise to algebraic curves on the surface, which vary in a complete linear system when the differential form varies. As this linear system does not make reference to any supplementary structure, for instance an embedding in an ambient space, it is called the *canonical system* $|K|$ of the surface. The canonical system is an essential tool for understanding the properties of the algebraic surface under consideration, both from the geometric and the transcendental viewpoint.

Following their geometric perspective, the previous Italian mathematicians wanted to construct the canonical system independently of the aforementioned differential forms. For this reason, they defined $|K|$ as a difference $|C' - C|$. Here $|C|$ is an arbitrary complete linear system of curves on the surface and $|C'|$ is its *adjoint system*.[1] By definition, the curves of the complete linear system $|C'|$ pass through the *base-points* of $|C|$ (the points which are common to all the curves of $|C|$) and intersect each such curve in a set of points belonging to its *canonical series* (see [32, Chaps. 16, 17 and 21]).

In dimension one, the canonical series of a smooth algebraic curve was defined transcendentally, as the complete linear series of the zero-loci of the everywhere holomorphic forms of degree one on the associated Riemann surface. However, there is also a geometric characterization of the canonical series, as the only linear

---

[1] This is both a reinterpretation and a generalization of the idea of the adjoint curve of a plane curve, explained in Chap. 22.

© Springer International Publishing Switzerland 2016
P. Popescu-Pampu, *What is the Genus?*, Lecture Notes in Mathematics 2162,
DOI 10.1007/978-3-319-42312-8_30

series of degree[2] $2p - 2$ and dimension[3] $p - 1$, where p denotes the genus of the curve (see Theorem 17.2).

This way of thinking brought some challenges. Namely, it was difficult to understand what happens when *the surface does not admit a non-zero holomorphic 2-form*. In this case, the canonical system $|K|$ seems to be empty. Nevertheless, it may occur that its double is non-empty! In terms of differential forms, it is possible that the surface does not have a non-zero holomorphic 2-form, but that there exists such a form *of weight* 2, that is, which may be written $a(x, y)(dx \wedge dy)^2$ in arbitrary local complex coordinates, with $a(x, y)$ holomorphic.

Here is what Castelnuovo and Enriques wrote with respect to this issue in [32, Chap. 30]:

> We spoke till now of only one invariant system situated on a given surface, namely the canonical system $|K|$. But one may readily introduce new invariant systems, as it is enough to consider the multiples $|K_i| = |iK|$ of the canonical system. It would not be worth paying attention to those new systems, if one could not find sometimes analogous systems on surfaces which do not possess a canonical system. [...]
>
> The *i-canonical* system provides us with new invariant characters of the surface; for instance, the number P_i of the *i*-canonical curves which are linearly independent, the genus of those curves, etc. For $i = 1$, $P_i = P_1$ gives us the geometric genus p_g of the surface; for $i = 2$ one has a character P_2 which could be called the *bigenus*, and which will play (as we will see) a fundamental role in the theory of rational surfaces; etc. Several of those new characters may be expressed in terms of the old ones; but not all of them will be realized in this way; this is what we will recognize concerning P_2.
>
> [...] A first very simple example is furnished by *the surface of the sixth order which passes twice through the six edges of a tetrahedron*, and consequently three times through its vertices; the surface has geometric genus $p_g = 0$, because there is no surface of the second order passing simply through the six edges. On the contrary, there exists a bi-adjoint surface of order 4, which is formed by the four faces of the tetrahedron; therefore, one has $P_2 = 1$.

This last example, discovered a little earlier by Enriques (and analyzed in detail in [71]), is historically *the first example of an algebraic surface which is not rational (that is, which is not birationally equivalent to a plane) but has vanishing geometric and arithmetic genera*. This is to be contrasted with the situation in dimension one, where the rationality of a curve is characterized by the vanishing of the genus.

The next natural step is *to find an analogous characterization of rational surfaces by invariant characters*. The previous example shows that the geometric and arithmetic genera do not suffice. The first advantage of the *plurigenera*, defined in the previous extract, is that they provide such supplementary invariants to solve this problem. But in fact they allow one to characterize many other types of surfaces, and even to achieve a genuine *classification*. We explain this in the next chapter.

---

[2]The *degree* of a linear series of points on a smooth projective algebraic curve is the degree of any divisor belonging to it.

[3]The *dimension* of a linear system—in particular of a linear series of points—is its dimension as a projective space.

Chapter 31
The Classification of Algebraic Surfaces

It was the classification of algebraic curves which served as a model for that of
surfaces, developed by Enriques with the help of Castelnuovo between 1890 and
1914. It was explained in detail in [33] and [71]. Let us begin with the way Enriques
presented the principle of the classification of curves in [70]:

> Let us recall first what Riemann, Clebsch, Brill and Nöther established about the classifica-
> tion of the curves
>
> $$f(x, y) = 0.$$
>
> An integer, which is called the *genus*, plays here the fundamental role. For any value of
> the genus p there is a family of curves, containing a continuous infinity of distinct classes,
> which is precisely ∞^{3p-3} for $p > 1$ (∞^1 for $p = 1$, ∞^0 for $p = 0$).
>   And it is essential to notice that this family is *irreducible*, so that in the classification of
> curves does not appear any other integer than the genus.

The irreducibility about which Enriques speaks means that there exists a family
of curves of genus p, parametrized by an irreducible algebraic variety, and which
contains at least one representative of every birational equivalence class of curves
of genus p. The notation ∞^k, common at that time, meant that one had a family
parametrized by a variety of dimension k.

Let us continue with the explanations given by Enriques in [69] concerning
algebraic surfaces[1]:

> For a rational surface one has: $p_g = p_a = 0$. But those necessary conditions are not enough
> to determine the class of rational surfaces. [...] For a rational surface one always has $P_2 =$
> 0, but conversely the bigenus does not necessarily vanish with the genus p_g or p_a. [...]
> However, Castelnuovo proved that the conditions of rationality of a surface amount to the
> simultaneous vanishing of the genus, geometric and numerical, and of the bigenus; they
> reduce by the way to $p_a = P_2 = 0$.

---

[1]At that time, the numerical genus was also called the *arithmetic genus*, and was denoted by p_a.

© Springer International Publishing Switzerland 2016   93
P. Popescu-Pampu, *What is the Genus?*, Lecture Notes in Mathematics 2162,
DOI 10.1007/978-3-319-42312-8_31

A more general problem, which contains that of rational surfaces, is the one of determining the surfaces $f(x, y, z) = 0$, representable in a parametric way by rational functions of a single parameter and algebraic functions of a different one, that is, of the surfaces which may be brought by a birational transformation to a cylinder[2] [...]. In order to solve this problem, the consideration of the bigenus no longer suffices; it is necessary to introduce also higher order genera [...]. Then, a remarkably simple result says that the necessary and sufficient conditions for a surface to be brought to the type of the cylinder are simply $P_4 = P_6 = 0$.

The genera of higher order, or plurigenera, also play a role in the general problem of classification of surfaces, beside the genera p_a and p_g [...]. But we are far from being able to define in this way all the families of surfaces, or of classes of surfaces, depending on arbitrary parameters or moduli. Regarding this matter, one has to expect complications which have no analog in the theory of curves. I limit myself to one example about this topic. Whereas the surfaces for which $p_a = P_3 = P_5 = \ldots = 0, P_2 = P_4 = \ldots = 1$, belong to the family of surfaces of sixth order which pass doubly through the edges of a tetrahedron, on the contrary, the surfaces all of whose genera are equal to $1, p_a = P_i = 1$, give rise to infinitely many distinct families, each one of them enclosing 19 moduli. The first of those families consists of the surfaces of the fourth order, the second being the surfaces of the sixth order passing doubly through a curve of the same order belonging to a quadric, etc.

The surfaces of these *infinitely many distinct families* were later called *K3 surfaces* by Weil (see [188]). The fact that they form an infinite number of families comes from another phenomenon which begins to express itself only from dimension 2 onwards: namely, *one may deform their complex analytic structures as little as one wants and get a surface which is no longer algebraic*!

But in order to contemplate such a phenomenon, one had to wait for the notion of abstract Riemann surfaces, explained by Weyl in [191] (see Chap. 23), to be extended to higher dimensions. Once this was done, Kodaira [122] made a classification of complex *analytic* surfaces analogous to the classification of complex *algebraic* ones. In this classification, the K3 surfaces form a single family. One may consult the lecture [15] by Beauville for a modern viewpoint on K3 surfaces and on this property, as well as the collective book [16] for much more details.

Figure 31.1 presents the classification table of algebraic surfaces, as it may be found at the end of Enriques' treatise [71] from 1949:

This classification is much more complicated than the one for curves, where it was enough to determine the value of a single invariant (the genus) in order to characterize an algebraic curve up to birational transformations and deformations. An analogous characterization is possible for several families of surfaces, provided one knows the values of several types of genera (and this was the main success of the classification undertaken by Castelnuovo–Enriques). Even today, there is no complete list of numerical invariants which would allow one to determine every algebraic surface up to birational transformations and deformations.

The reader who is curious to discover how the theory of algebraic surfaces was perceived in 1913 may read Baker's survey [12]. For more details about

---

[2] That is, here one aims to determine whether the given surface is birationally equivalent to a ruled surface.

Fig. 31.1 Enriques' classification of surfaces

the historical development of this classification, one may consult Gray's paper [85]. A modern presentation may be found in Reid's course [159]. One may find descriptions of the classification of not necessarily algebraic complex analytic surfaces in the books [79] of Friedman and Morgan and [13] of Barth et al.

Chapter 32
The Geometric Genus and the Newton Polyhedron

The various notions of genus which we have discussed up to this point do not tell us how to effectively compute the genera of an algebraic surface if we start from a defining equation $f(x, y, z) = 0$ in $\mathbb{C}^3$ (which determines its birational equivalence class).

It is this problem which was studied by Hodge in his paper [101] from 1930. He realized that the key object associated with f, which controls the *geometric genus*, is the *Newton polyhedron of f*. He defined it as the *convex hull in $\mathbb{R}^3$ of the triples of exponents $(a, b, c) \in \mathbb{N}^3$ of the monomials appearing in the reduced polynomial expression of f*. It is a compact convex polyhedron in $\mathbb{R}^3$ whose vertices are lattice points, that is, they have integer coordinates. Therefore, it contains only a finite number of lattice points, a fact which is essential in the sequel.

Newton had introduced the prototype for this construction in dimension two,[1] in order to explain how to express y as a fractionary power series of x (called *a Newton–Puiseux series*, see Chap. 13), if one knows a polynomial relation $f(x, y) = 0$, with $f(0, 0) = 0$. Such a *Newton polygon* is represented in the diagram 11 of Puiseux's paper [157], visible in Fig. 13.1.

Hodge proved that:

Provided that the coefficients are sufficiently general, the unit points within the Newton polyhedron lead to polynomials of the form $xyz\psi$, where $\psi = 0$ is of order $N - 4$ and is adjoint to the surface at the origin and at infinity.

If the surface $f(x, y, z) = 0$ has no singularities other than at the origin or at infinity, and if, further, the nature of these singularities is fully represented in the Newton diagram, the number of unit points which lie within the Newton polyhedron will be exactly equal to the geometric genus p_g of the surface. More generally, if the Newton diagram fully represents an isolated singularity, we can at once determine the effect of this singularity on the genus p_g.

---

[1]In fact, he had emphasized only a special edge of the polygon, without explicitly defining the polygon itself.

© Springer International Publishing Switzerland 2016
P. Popescu-Pampu, *What is the Genus?*, Lecture Notes in Mathematics 2162,
DOI 10.1007/978-3-319-42312-8_32

Briefly speaking, Hodge states in particular that:

Theorem 32.1 *The geometric genus of an algebraic surface in $\mathbb{C}^3$ defined by a polynomial f is equal to the number of lattice points situated in the interior of the Newton polyhedron of f, provided that the coefficients of f are sufficiently general.*

This theorem remained isolated inside mathematical research at that time, due to a lack of theoretical context in which to place it. In fact, the study of the relations between various invariants of hypersurfaces or of their singularities and their Newton polyhedra took off only after the creation of *toric geometry* in the 1970s.

Toric geometry is the branch of geometry which studies *toric varieties*. These are special types of irreducible algebraic varieties, containing an *algebraic torus* $(\mathbb{C}^*)^n$ as an open dense set and such that the action of the torus on itself by translation may be extended to an action on the whole toric variety.

What is essential for our context is that *any Newton polyhedron determines a canonical projective toric variety*. The basic method to study an affine hypersurface of $\mathbb{C}^n$ with a given Newton polyhedron is to take its closure inside this toric variety, instead of using the projective space $\mathbb{P}^n$. To see examples of the way in which this strategy works, one may look for instance at the generalization of Hodge's Theorem 32.1, obtained by Khovanskii in his 1978 paper [114] for complete intersections of arbitrary dimensions.

A modernized presentation of the previous theorem was also done in the spirit of toric geometry by Merle and Teissier in [130]. There, the authors studied in detail the notion of *adjunction* from a viewpoint adapted to the understanding of the classical writings about algebraic surfaces possessing singularities. Indeed, their essential motivation was to study the work [65] of Du Val regarding surface singularities *which do not affect the conditions of adjunction*. The next chapter is dedicated to those singularities.

To conclude this chapter, let us mention that around 1910, in his papers [66] and [67], Dumas had already begun to explore surface singularities using tools of a toric nature, long before toric geometry was developed.

Chapter 33
Singularities Which Do Not Affect the Genus

In his 1933 paper [65], Du Val said that an isolated singular point O of an algebraic surface S in $\mathbb{P}^3$ *does not affect the conditions of adjunction* if it does not impose conditions on the surfaces adjoint to S. That is, if those surfaces are not obliged to pass through the point O.[1]

In terms of an affine system of coordinates (x, y, z) centered at O, this means[2] that for any polynomial $q(x, y, z)$, the 2-form:

$$q(x, y, z) \frac{dx \wedge dy}{\partial f / \partial z}$$

is holomorphic in the neighborhood of the preimage of O on any *non-singular* projective surface Σ which maps birationally to S. That is, Σ is a resolution of singularities of S in the neighborhood of O, see the explanations below and Hironaka's theorem 29.1. Note that such a surface Σ plays the same role with respect to the complex algebraic surface S as the Riemann surface (which is *non-singular* by construction!) of an algebraic curve C does with respect to C.

As the geometric genus of a surface of degree n is equal by definition to the dimension of the space of homogeneous polynomials of degree $n - 4$ which define adjoint surfaces, one deduces that the presence of isolated singular points which do not affect the conditions of adjunction *does not affect the genus* either.

Du Val intended to characterize the surface singularities in $\mathbb{P}^3$ which are of this type. In modernized language, he proved the following theorem:

Theorem 33.1 *If an isolated singular point of a surface in $\mathbb{P}^3$ does not affect the conditions of adjunction, then it admits a resolution whose exceptional curve*

---

[1]Note that such singular points do not exist on plane algebraic curves, as shown by Noether's Theorem 22.2.

[2]Compare this with the explanations given for curves in Chap. 22.

© Springer International Publishing Switzerland 2016
P. Popescu-Pampu, *What is the Genus?*, Lecture Notes in Mathematics 2162,
DOI 10.1007/978-3-319-42312-8_33

consists of smooth rational curves of self-intersection -2. *Moreover, in this case, their dual graph is a tree having at most one ramification point, which is then of valency* 3, *and the length of the three maximal segments starting from this point, when ordered increasingly, form one of the following triples:* $(1, 1, n \geq 1)$, $(1, 2, 2)$, $(1, 2, 3)$, $(1, 2, 4)$.

What follows is a concise explanation of the notions involved in the previous statement. A *resolution* of the singular point $O \in S$ is a proper birational map $\rho : \Sigma \rightarrow S$ (defined everywhere), where Σ denotes a non-singular surface, such that ρ is an isomorphism above a punctured neighborhood of O (see also Theorem 29.1). In this case, the preimage of O under the map ρ is a curve E, called the *exceptional curve* of the resolution. Its *dual graph* has a set of vertices in bijection with the irreducible components of E, two vertices being joined by as many edges as the intersection number of the associated components. Moreover, every vertex is weighted by the genus and the self-intersection of the associated irreducible component in the smooth surface Σ.

Figure 33.1 represents in two different schematical ways the configurations of exceptional curves described by Theorem 33.1. It combines diagrams from Michael Artin's articles [6] and [7]. The second representation is that of dual graphs, explained in the previous paragraph. Such drawings of dual graphs became commonplace in the mathematical literature on algebraic surfaces and their singularities from around 1965.

In the articles [6] and [7], Artin generalized the results of Du Val. Namely, he classified the surface singularities which do not affect the arithmetic genus, *independently of their embedding dimension*. Artin called such singularities *rational*,

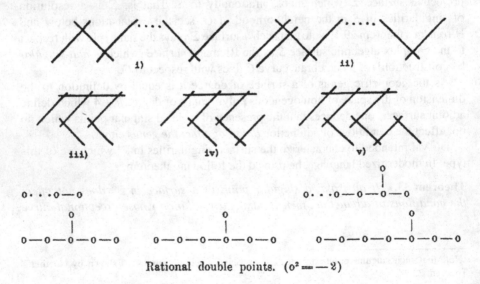

Rational double points. $(o^2 = -2)$

Fig. 33.1 Curve configurations and their dual graphs

as he characterized them through the vanishing of the arithmetic genus of a curve canonically associated with them. Moreover, he proved that *the singularities studied by Du Val are exactly the rational surface singularities of multiplicity* 2.

Theorem 33.1 allows us to see that if a surface of $\mathbb{P}^3$ has only singularities of the type described by Du Val (nowadays sometimes called *Du Val singularities*),[3] then it has the same geometric genus as a non-singular surface of the same degree. In fact, Brieskorn proved in [23] that, in this case, the resolution of the surface obtained by gluing the resolutions described in the theorem for the various singular points is *diffeomorphic* to the smooth surfaces of the same degree.

An account of the evolution of the understanding of Du Val singularities may be found in Brieskorn's article of recollections [24]. This class of singularities admits a lot of other characterizations. Those known in 1979 were gathered by Durfee in [68]. One may also consult the various papers of the volume [55], as well as the article [170] of Slodowy.

---

[3] They are alternatively called *Kleinian singularities* or *rational double points*. The first name refers to the fact that they are also the singularities of the quotients of $\mathbb{C}^2$ by finite subgroups of $SL(2, \mathbb{C})$, which were studied by Klein in [117]. The second one refers to Artin's characterization of Du Val singularities mentioned above.

Chapter 34
Hodge's Topological Interpretation of Genera

One may notice that the definitions of the various notions of genera of algebraic surfaces described in the previous chapters are all algebraic, that none is purely topological. We saw that the Italian geometers preferred to give such definitions starting from models of a birationally equivalence class situated in three-dimensional projective space, in which the curves of singularities were in general unavoidable. The question of the topological meaning of the genus deals instead with a smooth model, situated in a higher-dimensional space.

In his paper [102] of 1933, Hodge proved the following theorem:

Theorem 34.1 *The geometric genus of a smooth complex projective surface is a topological invariant.*

Hodge's proof depended in a crucial way on the extension to arbitrary dimensions made by Poincaré [148] (see Chap. 38) of Riemann's theory concerning cycles drawn on topological surfaces. It also depended on his own theory of harmonic forms on Riemannian varieties of arbitrary dimension (see Chap. 42).

This is the sign that it is time for us to look at higher dimensions. Before this, I want to explain the steps of Hodge's proof of Theorem 34.1, for the reader who knows a little bit about the theory of harmonic forms.

Hodge proved in fact that the *inertia index*[1] of the intersection form on the second cohomology group $H^2(V, \mathbb{R})$ of the surface V (see the explanations which follow Theorems 39.1 and 41.2) is equal to $(2p_g + 1, B_2 - 2p_g - 1)$, where p_g denotes the geometric genus and B_2 the second Betti number. As this intersection form is topologically invariant, one concludes that this is also the case for p_g.

More precisely, Hodge endowed the surface embedded in a projective space of arbitrary dimension with the Riemannian metric induced by a canonical choice of such a metric on the ambient projective space. Then, he decomposed the

---

[1]The *inertia index* of a real quadratic form is the pair consisting of the number of positive coefficients and the number of negative coefficients in a diagonalization of the form.

© Springer International Publishing Switzerland 2016
P. Popescu-Pampu, *What is the Genus?*, Lecture Notes in Mathematics 2162,
DOI 10.1007/978-3-319-42312-8_34

104   34 Hodge's Topological Interpretation of Genera

space of harmonic forms of degree 2, which is isomorphic to $H^2(V, \mathbb{R})$ using his Theorem 42.1, as the direct sum of the subspace of anti-self-dual forms (such that $*\phi = -\phi$, where $*$ denotes the Hodge involution, defined in Chap. 42) and that of self-dual forms (such that $*\phi = \phi$). He proved that the intersection form is positive definite on the first space and negative definite on the second (which is true on any oriented closed Riemannian manifold of dimension 4). Lastly (and this is specific to the smooth algebraic surfaces), he proved that the first space has dimension $2p_g + 1$.

Let us mention that nowadays the two subspaces of harmonic forms would be permuted. Their definition depends in fact on the sign conventions used in the definitions of Hodge's involution $*$ and on the orientation induced by the complex structure.

The irregularity $p_g - p_a$ (see the end of Chap. 28) is also a topological invariant. Indeed, as we mentioned previously, it is equal to the dimension of the complex vector space of forms of degree one, which is in turn equal to half of the first Betti number B_1 of the complex algebraic surface under scrutiny. This is a consequence of the fact that any form of degree one with complex values is cohomologous to a unique sum of a holomorphic form and of an antiholomorphic form (see also the comments which follow Theorem 42.2). This fact predates Hodge for the compact Riemann surfaces, and it also holds for smooth complex projective varieties of arbitrary dimension.

The geometric genus and the irregularity being topological invariants, it follows that the arithmetic genus has the same property.

Chapter 35
Comparison of Structures

The question of the topological invariance of the plurigenera resisted much longer. In fact, they are invariants of the differentiable structure, but not of the topological structure. More precisely;

Theorem 35.1 *If there exists an orientation preserving diffeomorphism between two smooth compact Kähler surfaces, then their plurigenera are equal. There exist smooth complex projective surfaces which are homeomorphic but which do not have the same plurigenera.*

The first statement, which answers affirmatively a conjecture of Van de Ven, was proved around 1996 by Brussee [26] and Friedman and Morgan [80], using the very recent theory of the equations of Seiberg–Witten.

The second statement was proved by Donaldson in his 1987 paper [64]. More precisely, he displayed two smooth complex projective surfaces, one of them being rational and the other one not, which are homeomorphic but not diffeomorphic. For this purpose he used the same techniques of "gauge theory" which had earlier allowed him to prove Theorem 35.3 below, as well as the fact that there exist topological 4-dimensional manifolds which do not admit a smooth structure, and others which admit several.

It was Milnor who had discovered in his 1956 paper [131] the first examples of smooth closed manifolds which are homeomorphic without being diffeomorphic. They were topologically spheres of dimension 7. By contrast, a little earlier, Moise had proved in [136] that a closed topological manifold of dimension 3 always admits a differentiable structure, which is unique up to diffeomorphisms.

Starting from Milnor's discovery of *"exotic spheres"*, the classification of differentiable structures on topological manifolds of dimension at least 5 progressed in an impressive way using tools from homotopy theory. But the situation in dimension 4 remained mysterious. In a certain sense, there was enough space in this dimension for the structural complexity to explode, compared to the situation in dimension 3, but still not enough for the problems to become flexible as in dimension at least 5.

© Springer International Publishing Switzerland 2016
P. Popescu-Pampu, *What is the Genus?*, Lecture Notes in Mathematics 2162,
DOI 10.1007/978-3-319-42312-8_35

Technically, this flexibility means that one may reduce the problems to questions of homotopy, which are not necessarily easy to solve, but which are in principle simpler than the original ones.

It was in 1982 that, almost simultaneously, Freedman and Donaldson showed that the problems of topological and smooth classification of manifolds of dimension 4 were radically different. More precisely, Freedman proved in [78] that (the notions used in this statement are explained in Chaps. 38 and 39):

Theorem 35.2 *Every simply connected closed topological manifold V of dimension 4 is determined, up to an ambiguity of cardinality 2, by its intersection form on the second homology group $H_2(V, \mathbb{Z})$. Moreover, any symmetric bilinear form with determinant ± 1 appears in this way.*

By contrast, Donaldson proved in [63] that the existence of a differentiable structure imposes serious restrictions on the intersection form of $H_2(V, \mathbb{Z})$:

Theorem 35.3 *A definite symmetric bilinear form of determinant ± 1 may be realized by a closed, oriented and simply connected differentiable manifold of dimension 4, if and only if its associated quadratic form may be written up to sign, in a convenient basis of $H_2(V, \mathbb{Z})$, as $x_1^2 + \cdots + x_n^2$.*

The reader who is eager to discover other applications of the gauge theoretical techniques of Donaldson or Seiberg and Witten to the study of the structure of complex surfaces may consult the rapid panoramic view of Barth, Hulek, Peters and Van de Ven [13, Chap. IX]. For the situation which preceded the Seiberg–Witten theory, a systematic presentation of the state of the art was done by Friedman and Morgan in [79]. An overview of the history of the classification of manifolds of dimension at most 4 was recently published by Milnor in [134].

Part III
Higher Dimensions

Chapter 36
Hilbert's Characteristic Function of a Module

We saw in Chap. 28 that, in order to define a birationally invariant notion of genus for algebraic surfaces, Clebsch, Cayley and Noether considered certain surfaces passing through their singular locus. Those surfaces are analogs of the adjoint curves of a plane algebraic curve, passing in a controlled way through their singular points (see Chap. 22).

In his 1890 paper [92], Hilbert began a systematic algebraic theory of the hypersurfaces in a projective space $\mathbb{P}^n$ of arbitrary dimension, which pass through a fixed algebraic subset V. The set of polynomials defining such hypersurfaces is the set of homogeneous polynomials of a *homogeneous ideal*[1] of the algebra A of polynomials in $n+1$ variables associated with the projective space $\mathbb{P}^n$. It was in this context that Hilbert proved:

Theorem 36.1 *Any ideal in a ring of polynomials is finitely generated.*

Nowadays, in memory of Emmy Noether's[2] 1921 work [145] on the rings which satisfy the same property, one says instead that *the rings of polynomials are Noetherian*. For Emmy Noether, it was important to build a general theory of ideals starting from *the abstract notion of commutative ring*, in order to unify the constructions based on ideals and modules in number theory and in algebraic geometry.

But in 1890, the year of publication of Hilbert's article [92], those constructions were almost absent from algebraic geometry, in particular from the theory of surfaces. In fact, still in 1935, in Zariski's treatise [195] surveying the theory of complex algebraic surfaces developed by the Italian school, the notion of ideal is not even mentioned! As Zariski was to explain later, it was during the writing of

---

[1] An *ideal* of a commutative ring is a subset which is closed under internal addition and exterior multiplication by elements of the ambient ring. An ideal of a polynomial ring is called *homogeneous* if it may be generated by homogeneous polynomials.
[2] She was Max Noether's daughter.

© Springer International Publishing Switzerland 2016
P. Popescu-Pampu, *What is the Genus?*, Lecture Notes in Mathematics 2162,
DOI 10.1007/978-3-319-42312-8_36

this book that he became convinced of the necessity to fortify the bases of algebraic geometry and to make its proofs more rigorous. Reflecting upon this problem, he realized that the theory of ideals in commutative rings was perfectly adapted to this double aim.

For instance, in [196], one of the first articles in which Zariski used the theory of ideals in the study of complex algebraic surfaces, he re-examined the notion of a linear system of curves passing through assigned base-points of a smooth surface S. Certain of those points could be *infinitely near* (that is, situated on curves which are created by successive blow ups, as explained in Chap. 22). He showed how to encode the assignation of a finite set of infinitely near points of an actual point O of S by an ideal of the ring of rational functions on S which do not have poles at O.[3] Namely, a curve passes through those points if and only if a function defining it is contained in the ideal.

From that time on, it became important to consider not only ideals of functions which vanish on a subvariety of a given variety, but *all* ideals: they may be thought of as defining conditions for passing through base-loci, perhaps infinitely near to actual subvarieties of the ambient variety.

Let us mention that the source of the notion of *ideal* was Kummer's notion of *ideal number*, defined in an arithmetic context. It was Dedekind who decanted from it the notion of *ideal* in the modern mathematical sense (see [54]). The reader eager to discover the way Kummer wrote about ideal numbers may consult [154].

As mentioned before, the ideals I defined by subsets of a projective space, which were those considered by Hilbert in [92], have the property of being *homogeneous*. Therefore, they may be seen as direct sums, indexed by $d \in \mathbb{N}$, of the complex vector subspaces I_d consisting of their homogeneous elements of degree d. The quotient $\mathbb{C}$-algebra A/I is thus also naturally graded, I and A/I being the fundamental examples of *graded A-modules*[4] considered by Hilbert. He defined as a measure of complexity for such a module *its characteristic function* $\chi : \mathbb{N} \to \mathbb{N}$, which associates to every $d \in \mathbb{N}$ the dimension of its degree d vector subspace.

Geometrically, if I is the ideal associated with a subvariety V of the projective space, the characteristic function of A/I associates to each degree d the dimension of the linear system of sections of V defined by the hypersurfaces of degree d, augmented by 1.[5]

Nowadays the characteristic function is called *Hilbert's function* of the subvariety V of $\mathbb{P}^n$ or of the graded $\mathbb{C}$-algebra A/I. The main theorem in Hilbert's paper [92] regarding this function is:

---

[3]This ring is called the *local ring* of the algebraic surface S at the point O.

[4]Recall that *modules* are to rings what vector spaces are to fields. For instance, the ideals of a ring A are precisely its subsets which are A-modules.

[5]The difference of 1 comes from the facts that proportional polynomials define the same member of the linear system and that the dimension of a vector space exceeds by 1 the dimension of the associated projective space.

Theorem 36.2 *Let χ be the characteristic function of a graded A-module M of finite type. Then $\chi(d)$ becomes a polynomial in d, for d large enough. When $M = A/I$, I being the ideal generated by the homogeneous polynomials which vanish on the algebraic subset V of $\mathbb{P}^n$, the degree of this polynomial equals the dimension of V.*

The *dimension* of an algebraic variety is therefore readable from the previous Hilbert function associated with any embedding of the variety into a projective space. This leads to the expectation that one might also extract from it other invariants which are independent of the embedding. This expectation was stated by Hilbert when he discussed possible generalizations of the notion of genus, on pages 519–520 of [92, Chap. IV][6]:

Consider a curve or a configuration of curves and points in 3-dimensional space. By the theorem proved in Chapter I, one may always find a finite number of surfaces

$$F_1 = 0, F_2 = 0, \ldots, F_n = 0$$

containing that configuration such that any other surface which contains the configuration is given by an equation of the form

$$A_1 F_1 + A_2 F_2 + \cdots + A_m F_m = 0.$$

Those considerations show that to any algebraic configuration corresponds a module $(F_1, F_2, \ldots, F_m)$ and through it, the characteristic function $\chi(R)$. This last function describes how many independent conditions must be satisfied by a surface of given degree which exceeds a certain bound, in order to contain the given configuration. In this way, the characteristic function of a space curve without double points, of order R and genus p has the value

$$\chi(R) = -p + 1 + rR.$$

[...] In what concerns the generalization of those considerations to spaces of higher dimensions, the next result seems important. Let us fix an algebraic configuration in a space of arbitrary dimension and the characteristic function of the module corresponding to the configuration, given by:

$$\chi(R) = \chi_0 + \chi_1 \binom{R}{1} + \chi_2 \binom{R}{2} + \cdots + \chi_d \binom{R}{d}.$$

Then, the degree of this characteristic function gives the dimension and the coefficient χ_d gives the order [that is, its degree as a subspace of the projective space] of the algebraic configuration, while the other coefficients $\chi_0, \chi_1, \ldots, \chi_{d-1}$ are in tight relation to the genera of the configuration defined and discussed by Mr. Noether. The general proof of this fact depends on the passage from $n - 1$ variables to n. As may be seen, the theorems which we gave in the case of curves in 3-dimensional space are indeed confirmed.

Note that this technique does not impose any constraint on the possible singular points of the variety V. Therefore, it opens the door to a definition of the genus in any dimension, independently of their nature. It was Severi who proposed such a definition, following the trail left by Hilbert. The next chapter is dedicated to this definition.

---

[6]I am grateful to Walter Neumann for having translated the following quotation into English.

Chapter 37
Severi and His Genera in Arbitrary Dimension

In his 1909 paper [167], Severi formulated a definition of *arithmetic genus* valid in arbitrary dimension. In order to understand the following extract from that paper, one has to know that Severi used the term *postulation* for the polynomial $v(l)$ equal to the Hilbert function $\chi(l)$ whenever $l \in \mathbb{N}$ is sufficiently big.[1] Note also that Severi used a different basis to Hilbert's for the $\mathbb{Z}$-module of polynomials of degree at most d which take only integer values at the integers.

In S_r [a projective space of dimension r] let V_d be an algebraic variety, irreducible or not, without multiple varieties of its dimension, and let

$$v(l) = k_0 \binom{l+d}{d} + k_1 \binom{l+d-1}{d-1} + \cdots + $$
$$+ k_{d-1}(l+1) + k_d$$

be the formula of the postulation associated to it.

The meaning of the coefficients $k_0, k_1, \ldots, k_d$ is well-known when $d = 0, 1, 2$. In the case $d = 0$, the coefficient k_0 is nothing else than the number of points of the group; in the case of an irreducible curve without singularities, of order $p_0 + 1$ and of genus p_1, one has (Cayley, Noether, Castelnuovo):

$$k_0 = p_0 + 1, \quad k_1 = -(p_0 + p_1);$$

in the case of a surface, without singularities, of arithmetic genus p_2, whose hyperplane section has the order $p_0 + 1$ and the genus p_1, one has (Severi):

$$k_0 = p_0 + 1, \quad k_1 = -(p_0 + p_1), \quad k_2 = p_1 + p_2.$$

Therefore in the case of a curve

$$p_1 = -(k_0 + k_1 - 1);$$

---

[1]Nowadays $v(l)$ is called *Hilbert's polynomial*.

© Springer International Publishing Switzerland 2016
P. Popescu-Pampu, *What is the Genus?*, Lecture Notes in Mathematics 2162,
DOI 10.1007/978-3-319-42312-8_37

and in the case of a surface

$$p_2 = k_0 + k_1 + k_2 - 1.$$

That being said, for an arbitrary V_d the idea to define the *virtual arithmetic genus* using the formula

$$p_d = (-1)^d(k_0 + k_1 + \cdots + k_d - 1)$$

comes naturally. After that, *the postulation formula of any variety V_d may be written as*

$$\begin{cases} v(l) = (p_0 + 1)\binom{l+d}{d} - (p_0 + p_1)\binom{l+d-1}{d-1} + \cdots \\ \cdots + (-1)^{d-1}(p_{d-2} + p_{d-1})(l+1) + (-1)^d(p_{d-1} + p_d), \end{cases}$$

where $p_0, p_1, p_2, \ldots, p_d$ are the virtual arithmetic genera of the sections of V by spaces $S_{r-d}, S_{r-d+1}, S_{r-d+2}, \ldots, S_r$, respectively.

When the [variety] V_d is irreducible and without singularities, the number p_d will be called the *effective arithmetic genus* or simply the *arithmetic genus*.

But the [last relation] would remain a useless formal transform of [the first one] unless one establishes that, in the case of irreducible varieties, the arithmetic genera are invariant characters relative to birational transformations.

Thus Severi formulated the following definition of the arithmetic genus of a *smooth* variety of dimension d, embedded in a projective space, in terms of its Hilbert polynomial $v(l)$:

$$p_a = (-1)^d(v(0) - 1). \tag{37.1}$$

Severi remarked that it is not clear that this notion is intrinsic, that is, independent of the chosen embedding of the variety into a projective space. However, a great advantage of his definition is that it applies to any subvariety of a projective space, even singular or with multiple components. It is important to have such a notion because, when one considers a family of subvarieties of $\mathbb{P}^n$, special members of it may be singular.[2] One may ask how the arithmetic genus changes when one passes to such a singular member of the family. Well, if the family is simply-behaved geometrically, then it remains constant. Indeed, one has the following theorem (see Hartshorne's book [90, Chap. III.9] for a proof):

Theorem 37.1 *A one-parameter family $f : X \to T$ of subvarieties of a projective space $\mathbb{P}^n$ is flat (that is, each one of its members $f^{-1}(t)$ is the algebraic limit of the other members, or, said differently, the closure of $X \setminus f^{-1}(t)$ in $\mathbb{P}^n \times T$ is equal to X) if and only if the Hilbert polynomials of the members $f^{-1}(t)$ of the family are constant.*

In particular, *the arithmetic genus is invariant in such a family.*

---

[2]One may think simply about the family of all the conics of $\mathbb{P}^2$, which contains unions of two lines, as well as double lines.

Let us also mention that if C is a curve, possibly singular or with multiple components, but which is contained in a *smooth* projective surface S (it is therefore an effective divisor on this surface), then one may compute its arithmetic genus *numerically* on the surface S, that is, in terms of intersection numbers (see again Hartshorne's book [90, Exercise V.1.3]):

Theorem 37.2 *If S is a smooth projective surface with canonical class K and C is an algebraic curve contained in S, one has the following adjunction formula:*

$$p_a(C) = 1 + \frac{1}{2}C \cdot (C + K).$$

The birational invariance of the arithmetic genus of *smooth* projective varieties was not the only problem left open by Severi in this article. On its last page, he conjectured that for a smooth variety, the arithmetic genus he defined is equal to the alternating sum:

$$i_d - i_{d-1} + i_{d-2} - \cdots + (-1)^{d-1} i_1,$$

where i_k denotes "*the number of finite k-tuple integrals*", that is, the dimension of the space of rational forms of degree k without poles. This conjecture would solve the problem of showing that the arithmetic genus is intrinsic. Indeed, it is easy to see by a computation done in local coordinates that a birational transformation between smooth varieties sends every rational differential form without poles to a form of the same type.

Severi's conjecture was to be proved only in 1952 by Kodaira [120]. The reader curious to learn how, just before the appearance of Kodaira's paper, Severi saw the evolution of the problems initiated by his article of 1909, may consult his 1951 paper [168].

Kodaira's proof used in an essential way the developments of the first half of the twentieth century concerning the topology of varieties of arbitrary dimension initiated in Poincaré's paper [148]. Let us discuss now those works of Poincaré.

Chapter 38
Poincaré and Analysis Situs

The notions of genus which we have studied until now for algebraic surfaces, as well as those formulated in all dimensions by Severi, are defined by algebraic means. None of them generalizes Riemann's topological viewpoint, based on curves drawn on differentiable surfaces (see Chap. 14).

In his 1895 paper [148], Poincaré intended to rectify this situation and to extend the topological tools used in dimensions 2 and 3 to arbitrary dimensions. As he explained in the introduction of his paper, he was partially motivated by questions about the geometry of algebraic surfaces:

> The use of figures [...] aims to teach us certain relations between the objects of our study, which relations are those dealt with by a branch of Geometry which is called *Analysis situs*, and which describes the relative situation of points, lines and surfaces, without any consideration of their sizes.
>
> There are relations of the same nature between the beings of hyperspace; there exists therefore an *Analysis situs* with more than three dimensions, as was shown by Riemann and Betti.
>
> This science will enable us to know these kinds of relations, although this knowledge cannot be intuitive any more, as our senses fail. It will therefore render us sometimes several of the services which we usually ask of the figures of Geometry.
>
> I will give only three examples.
>
> The classification of algebraic curves into genera rests, following Riemann, on the classification of real closed surfaces, done according to the viewpoint of *Analysis situs*. An immediate induction makes us understand that the classification of algebraic surfaces and the theory of their birational transformations are intimately related to the classification of real closed surfaces of the space of five dimensions, from the viewpoint of *Analysis situs*. Mr. Picard, in a work crowned by the Academy of Sciences, already insisted on this point.
>
> On the other hand, in a series of works appearing in the *Journal de Liouville*, and called: *On the curves defined by differential equations*, I used the usual *Analysis situs* in three dimensions for the study of differential equations. The same researches were pursued by Mr. Walther Dyck. One easily sees that the generalized *Analysis situs* would allow us to treat even the equations of higher order and, in particular, those of celestial mechanics.
>
> Mr. Jordan determined analytically the groups of finite order contained in the linear group in *n* variables. Mr. Klein had solved before, by a geometric method of rare elegance, the same problem for the linear group in two variables. Wouldn't it be possible to extend the

© Springer International Publishing Switzerland 2016
P. Popescu-Pampu, *What is the Genus?*, Lecture Notes in Mathematics 2162,
DOI 10.1007/978-3-319-42312-8_38

method of Mr. Klein to the group in n variables or even to an *arbitrary continuous group*? I have been unable to achieve this, but I have thought a lot about the question and it seems to me that the solution must depend on a problem of *Analysis situs* and that the generalization of the famous theorem of Euler about polyhedra must play a role in it.

We see that Poincaré wrote explicitly about the problem of finding a birationally invariant notion of genus for algebraic surfaces. He obviously had in mind all the definitions given by Riemann for algebraic curves. This made him get interested in particular in the multiple integrals which are everywhere finite, carried by an algebraic surface.[1]

As his aim was to create a framework which may be applied to problems of very diverse types (we saw in the previous quotation that he wrote about *algebraic geometry*, *differential equations* and *group theory*), he developed his theory for smooth real manifolds which are not necessarily algebraic. They were still not defined abstractly, but as submanifolds of the spaces $\mathbb{R}^n$, described either implicitly, or parametrically (see [148, Chaps. 1–4]).

Riemann considered closed curves drawn on his surfaces. Analogously, Poincaré considered closed submanifolds of a given manifold, that is, compact submanifolds without boundary. Riemann considered several curves at the same time, that do not form the complete boundary of a surface. As discussed in Chap. 14, using them he defined the notion of *connection order*. His proof of the topological invariance of this notion was based on the fact that, by adding a supplementary closed curve, the new set of curves then bounds a surface. It is this notion of *being the boundary of a surface* which Poincaré generalized to all dimensions (see [148, Chap. 5]):

Let us consider a manifold[2] V of p dimensions; let W be a manifold with q dimensions ($q \leq p$) which is part of V. Assume that the complete boundary of W consists of λ continuous manifolds with q dimensions

$$v_1, v_2, \ldots, v_\lambda.$$

We will express this fact by the notation

$$v_1 + v_2 + \ldots + v_\lambda \sim 0.$$

More generally the notation

$$k_1 v_1 + k_2 v_2 \sim k_3 v_3 + k_4 v_4,$$

---

[1]That is, in the algebraic differential forms without poles; see Chap. 40 for a description of the passage of the formulation in terms of integrals to the formulation in terms of forms.

[2]Poincaré used the term "*variété*", which may be translated in English to either "manifold" or "variety". Nowadays, the second term is generally reserved for algebraic or analytic *possibly singular* spaces. As Poincaré worked with *differentiable* objects, for which the term "variety" is much less used, we chose the first term for our translation. Nevertheless, it is very difficult to make all his ideas work without also allowing singular such "manifolds". This is one of the reasons why "singular homology" was later developed in order to construct a completely rigorous incarnation of his ideas.

where the k's are integers and the v's manifolds with $q-1$ dimensions, will mean that there exists a manifold W of dimension q which is part of V and whose complete boundary will be composed of k_1 manifolds slightly different from v_1, of k_2 manifolds slightly different from v_2, of k_3 manifolds slightly different from the opposite manifold [that is, with the opposite orientation] of v_3 and of k_4 manifolds slightly different from the opposite manifold of v_4.

The relations of this form may be called *homologies*.

Homologies may be combined as ordinary equations.

These definitions allowed Poincaré to extend the definition given by Riemann of the connection order of a surface to manifolds of arbitrary dimensions (see [148, Chap. 6]):

We will say that the manifolds [...] of a same number of dimensions and being part of V, are *linearly independent* if they are not related by any homology with integral coefficients.

If there exist $P_m - 1$ *closed* manifolds of m dimensions being part of V and linearly independent and if there exist only $P_m - 1$ of them, we will say that the order of connection of V relative to the manifolds of m dimensions is equal to P_m.

In this way are associated $m - 1$ numbers to a manifold V of dimension m [...].

I will call them in the sequel the *Betti numbers*.

Later on, as we will explain in the sequel, it is $P_m - 1$ which was called the m-th *Betti number* of V. Even though Poincaré does not cite any article of Betti, this probably refers to the 1871 paper [20], in which Betti already tried to extend the topological ideas of Riemann concerning surfaces to arbitrary dimensions. In fact, Betti had discussed this problem with Riemann himself, as may be seen from the letters he had exchanged with Tardy, translated into English by Weil in [189], and analysed by Pont in [152, Chap. II.3.2].

In the same article, Poincaré also dealt with the problem of the multivaluedness of functions in the manner of Cauchy and Puiseux (see Chaps. 12 and 13), which led him to introduce the notion of *fundamental group* of a manifold:

Let be given [...] λ functions [F] of the coordinates [...] of a point M of a manifold [...]. I do not assume that the functions F are uniform. When the point M, starting from its initial position M_0, will come back to this position, after having traveled along an arbitrary path, it is possible that the functions F do not come back to their primitive values.

[...] *The set of all the substitutions which the functions F will undergo in this way, when the point M will describe all the closed contours which may be drawn on the manifold V starting from the initial point M_0, will obviously form a group, which I shall call g.* [...]

If now $M_0AM_1, M_0BM_1, M_0CM_1$ are three different paths drawn on V and going from M_0 to M_1, I will write by convention

$$M_0AM_1CM_0 \equiv M_0AM_1BM_0 + M_0BM_1CM_0.$$

[...] We are thus led to consider relations of the form

$$k_1C_1 + k_2C_2 \equiv k_3C_3 + k_4C_4,$$

in which the k's are integers and the C's are closed contours drawn on V and starting from M_0. Those relations, which I will call *equivalences*, are similar to the homologies which I studied above. They differ from them:

1°   Because, in the homologies, the contours may start from an arbitrary initial point;
2°   Because, in the homologies, one has the right to switch the order of the terms of a sum.

[...] it is clear that one may imagine a group G satisfying the following conditions:

1°   To each closed contour $M_0 B M_0$ will correspond a substitution S of the group;
2°   The necessary and sufficient condition for S to reduce to the identical substitution is that [the closed contour be homotopic to a constant contour];
3°   If S and S' correspond to the contours C and C' and if $C'' \equiv C + C'$, the substitution corresponding to C'' will be SS'.

The group G will be called the *fundamental group* of the manifold V.

Let us compare it to the group g of the substitutions undergone by the functions F.

The group g will be isomorphic to G [that is, following the language of the time, it is the image of G by a morphism of groups].

The isomorphism may be holoedric [that is, an isomorphism according to our present conventions].

It may be meriedric [that is, it may have a non-trivial kernel] if a closed contour $M_0 B M_0$ [which is not homotopic to a constant contour] brings back the functions F to their primitive values.

Nowadays, one defines the *fundamental group* of a topological space which is arcwise connected, as the set of homotopy classes of loops based at a fixed point of it, the product of two classes being obtained by traveling successively along representatives of those classes, and by taking the homotopy class of the result.

From the viewpoint of the theory of multivalued functions living on them, the simplest manifolds are those whose fundamental group is trivial. Such manifolds were to be called *simply connected*, even if for Poincaré this expression meant rather *being homeomorphic to a ball or a sphere*.

Poincaré wrote about "substitutions" because in his time all the known groups were groups of permutation (of roots of polynomials in Galois theory) or of substitutions (that is, of changes of variables). But in fact the group defined by him is abstract, a priori it does not transform anything (even if, a posteriori, it acts on the *universal covering*, that is, on the only non-ramified simply-connected covering of the initial manifold).

This was *the first definition of an invariant of a geometric object which was not a number* (as were all the notions of *genus* seen previously, that of the *connection order* of a surface or those of *Betti numbers*).

It became clear during the twentieth century that it was very important not to directly associate numbers as invariants of *geometric* objects, but intermediate *algebraic* objects, of which those numbers are in turn invariants. For instance, as explained in the next chapter, the Betti numbers were to be seen as dimensions of *homology groups* (seemingly following a suggestion of Emmy Noether, as stated by Dieudonné in [61], an article to be read in parallel with the article [129] of Mac Lane, who tempered the statements of Dieudonné; see also Basbois' paper [14]).

Chapter 39
The Homology and Cohomology Theories

It is not clear from the initial definitions of Poincaré that the Betti numbers of a compact manifold are finite or that its fundamental group is finitely generated. In order to address in particular these two issues, Poincaré proposed in 1899, in the sequel [149] to his article [148], an alternative definition of the Betti numbers, which used a polyhedral decomposition of the given manifold:

> I consider therefore in the sequel a closed manifold V, but in order to compute its Betti numbers, I assume it subdivided into smaller manifolds, such as to form a polyhedron [...].

When choosing this viewpoint, he was in fact inspired by his previous definition, formulated in [148], of the *characteristic* (nowadays called the *Euler–Poincaré characteristic*, see Chap. 47) of an arbitrary manifold, which already used a polyhedral decomposition.

Said differently, by using a more recent expression, Poincaré *discretized the problem*, as one does nowadays in numerical analysis when, in order to compute a solution of a partial differential equation in an approximate way, one uses the vertices of an adapted triangulation of the domain where the equation is defined.

In his case, he discretized in two steps the notion of a *submanifold with boundary* of the given manifold V. First, he endowed each face of the chosen polyhedral decomposition of V with a fixed orientation. Then, for each fixed k, he considered as k-dimensional "submanifolds" the arbitrary sums of k-dimensional oriented faces of the polyhedral decomposition.

In modern language, those "submanifolds" are called *chains of dimension k*. Such chains form the elements of the free abelian group C_k generated by the oriented faces of dimension k. The *boundary* of a chain is the unique linear map which associates to each face of dimension k the formal sum of its oriented faces of dimension $k-1$, each one being endowed with a canonically chosen sign.

Among the chains, one distinguishes the *cycles*, whose boundaries vanish. They form a subgroup of C_k. The *k-th homology group of V with integral coefficients* is by definition the quotient of the group of cycles of dimension k by the subgroup of

© Springer International Publishing Switzerland 2016
P. Popescu-Pampu, *What is the Genus?*, Lecture Notes in Mathematics 2162,
DOI 10.1007/978-3-319-42312-8_39

boundaries of chains of dimension $k + 1$. One denotes it by:

$$H_k(V; \mathbb{Z}).$$

According to the modern terminology, its rank is by definition *the k-th Betti number* of V. Note the shift by 1 compared to Poincaré's definition. The group C_k is by construction of finite rank, therefore the discretized k-th Betti number is also finite.

The definition may be repeated by considering chains with coefficients in other rings A. One obtains in this way the *homology groups of V with coefficients in the ring A*, denoted by:

$$H_k(V; A).$$

It is then natural to ask the following questions:

- Is it true that every manifold possesses such a polyhedral structure?
- Is it true that two distinct polyhedral structures on the same manifold always give the same Betti numbers?

Both questions were later answered affirmatively for *smooth* manifolds. But the previous theory was extended to all topological spaces, even to those which do not admit polyhedral structures. To this end, several alternative definitions of homology groups were elaborated, giving the same groups for varieties, but applicable to more general spaces. For the history of these constructions, one may consult [62] and [108].

This research inspired the formulation of conditions which allow one to say that some group is in fact a *homology group*. The fundamental object behind the different contexts in which people agreed to speak about "homology" was found to be a sequence of abelian groups and morphisms between consecutive groups:

$$\cdots \to C_{k+1} \to C_k \to C_{k-1} \to \cdots, \tag{39.1}$$

such that the composition of two consecutive morphisms vanishes. In the case of Poincaré's polyhedral construction, the morphism $C_k \to C_{k-1}$ associates to every k-dimensional chain its boundary. This allows one to formulate the condition on the composition of consecutive morphisms as:

$$\textit{The boundary of a boundary vanishes.} \tag{39.2}$$

Nowadays, one says that such a sequence of groups and morphisms between them is a *chain complex*. Its *homology* is then the sequence of groups:

$$H_k = \ker(C_k \to C_{k-1})/\operatorname{im}(C_{k+1} \to C_k).$$

Let us explain one of the reasons why it is beneficial to work with homology groups instead of their numerical invariants.

Any closed and oriented m-dimensional submanifold W of V gets represented in $H_m(V, \mathbb{Z})$ by its *fundamental class*, which may be defined by choosing a polyhedral decomposition of V such that W is a union of faces, and by taking the sum of those faces after orienting them using the given orientation of W. In particular, if V has dimension n, it admits a fundamental class in $H_n(V, \mathbb{Z})$. This allows one in a certain way to *linearize* the problems concerning the global relative positions of the submanifolds of V.

A first example of this linearization process is given by *Poincaré's duality theorem*, which he stated in the following way (see [148, Sect. 9] and the explanations given in [156]):

[...] for a closed manifold, the Betti numbers situated at equal distances from the extremities are equal.

In fact, there exists a canonical *bilinear intersection form*:

$$H_k(V, \mathbb{Z}) \times H_{n-k}(V, \mathbb{Z}) \to \mathbb{Z}$$

for every closed and oriented manifold of dimension n. One may describe it geometrically in the following way. The product of two homology classes may be obtained by choosing representatives of those two classes which intersect transversally, and by counting algebraically their number of intersection points. Each point has an attached sign, defined according to the orientations at it of the ambient manifold and of the representatives of the homology classes.

Modulo the torsion subgroups, this intersection form establishes a duality between the two groups $H_k(V, \mathbb{Z})$ and $H_{n-k}(V, \mathbb{Z})$, which shows that they have the same rank. In particular, choosing to work with real coefficients eliminates the torsion and one gets:

Theorem 39.1 *If V is an oriented closed manifold of dimension n, then one has a canonical isomorphism:*

$$H_{n-k}(V, \mathbb{R}) \simeq H_k(V, \mathbb{R})^*.$$

If n is even and $k = n/2$, one gets a non-degenerate intersection form on the real vector space $H_k(V, \mathbb{R})$. It is symmetric when $n/2$ is *even* and it is skew-symmetric when $n/2$ is *odd*.

In the first case, one obtains a new topological invariant of V, namely the *inertia index* of this intersection form, which determines the form up to isomorphism.

In the second case, one does not get a new invariant from this skew-symmetric form, as all skew-symmetric non-degenerate forms defined on a fixed real vector space are isomorphic. But as such forms exist only on real vector spaces of *even* dimension, one gets the following generalization of the fact, proved by Riemann, that the first Betti number of a closed Riemann surface is even (see Chap. 14, where we explained in particular that Riemann saw the genus p of the surface as half this even number):

Theorem 39.2 *If V is an orientable closed manifold of dimension* $4m + 2$, *then its middle Betti number* B_{2m+1} *is even.*

In Theorem 42.2 of Chap. 42, we will see another extension of Riemann's result to *Kähler* manifolds.

Nowadays, each time one constructs a chain complex in a sufficiently general context, one says that is defined a *homology theory*. These theories flourished in the twentieth century, having as a model this first example introduced by Poincaré. But one had to wait until the 1930s to see the appearance of a homology theory of a rather different nature, *de Rham cohomology*, constructed by starting from *differential forms*. Before describing it, let us first consider those forms more carefully.

The reader who wants to have much more details about Poincaré's works on *Analysis situs* may consult the website [53] by de Saint-Gervais.

Chapter 40
Elie Cartan and Differential Forms

The curves drawn on Riemann surfaces are supports of integration for the forms of degree 1. Similarly, the oriented submanifolds of dimension k of a given manifold V serve as supports for the computation of integrals of differential forms of degree k, as explained by Poincaré in [148, Chap. 7].

In fact, when Poincaré was writing that article, the notion of a differential form of arbitrary degree had not been formulated yet. It was only introduced in 1899 in Elie Cartan's paper [28]. But the custom to write about the *number of k-tuple independent integrals* rather than of the *dimension of the vector space of holomorphic forms of degree k* was to last for a while.

Cartan studied the notion of a *differential form* of arbitrary degree under the name of a *differential expression*. Here is the way in which he explained this notion in the introduction to his article:

> The present work constitutes a presentation of the problem of Pfaff[1] founded on the consideration of certain symbolic differential expressions, integral and homogeneous relative to the differentials of n variables, the coefficients being arbitrary functions of those variables. Those expressions may be submitted to the usual rules of calculus, provided one does not change the order of the differentials in a product. The calculus of those quantities is, in short, that of differential expressions which are placed under a multiple integral sign. This calculus also presents many analogies with the calculus of Grassmann; it is by the way identical to the geometric calculus used by Mr. Burali-Forti in a recent Book.

The rules of calculus used for *"differential expressions which are placed under a multiple integral sign"* were slowly developed (see the historical articles [111–113] of Katz). For instance,[2] Euler explained in [74, page 87] that if one considers the isometry $(t, v) \rightarrow (x, y)$ of the Euclidean plane given by the following formulae (in

---

[1] That is, the problem of solving a system of equations of the form $\omega_1 = \cdots = \omega_n = 0$, where the ω_i are forms of degree 1, also called *Pfaff forms*.

[2] I am taking this example from [112].

© Springer International Publishing Switzerland 2016
P. Popescu-Pampu, *What is the Genus?*, Lecture Notes in Mathematics 2162,
DOI 10.1007/978-3-319-42312-8_40

which $|m| \le 1$):

$$\begin{cases} x = a + mt + v\sqrt{1 - m^2} \\ y = b + t\sqrt{1 - m^2} - mv \end{cases},$$

then, by multiplying the differentials:

$$dx = m\, dt + dv\, \sqrt{1 - m^2}, \quad dy = dt\sqrt{1 - m^2} - m\, dv,$$

one gets:

$$dx\, dy = m\sqrt{1 - m^2}\, dt^2 + (1 - 2m^2)dt\, dv - m\sqrt{1 - m^2}\, dv^2.$$

This formula cannot be used for the computation of a double integral of the form:

$$\int \int f(x, y)\, dx\, dy$$

by the previous change of variables.

In fact, Euler did not find a convenient rule for performing a change of variables in a multiple integral. Such rules seem to have emerged only in the 1890s. For instance, in his 1896 article [27, page 143], Cartan felt obliged to explain in a footnote how to make a change of variables in a double integral:

[...] if one sets

$$u = f(\lambda, \mu, v, \dots, \xi), \quad v = \phi(\lambda, \mu, v, \dots, \xi),$$
$$du = A\, d\lambda + B\, d\mu + \dots + C\, d\xi, \quad dv = A'\, d\lambda + B'\, d\mu + \dots + C'\, d\xi,$$

one gets

$$du\, dv = (A\, d\lambda + B\, d\mu + \dots + C\, d\xi)(A'\, d\lambda + B'\, d\mu + \dots + C'\, d\xi),$$

provided one replaces, in the product of the second member performed *without altering the order of the differentials* of each partial product, $d\lambda\, d\lambda$ by 0 and $d\lambda\, d\mu = -d\mu\, d\lambda$ [...].

The key rule is that the products of forms of degree 1 are *skew-symmetric*. The special notation $\omega \wedge \phi$ for the product of two differential forms ω and ϕ, instead of the simple juxtaposition $\omega\phi$, was introduced later to emphasize the fact that those products are not commutative, in contrast to Euler's assumption.

Note that, if one now applies the rule of antisymmetry to the computation of Euler's example, one finds the correct area-preserving property of plane isometries!

In the subsequent paper [28] of 1899, Cartan defined the differential forms in a *purely symbolical* manner:

Given n variables $x_1, x_2, \dots, x_n$, let us consider purely symbolical expressions ω, obtainable from the n differentials $dx_1, dx_2, \dots, dx_n$ and from certain coefficients which are functions of $x_1, x_2, \dots, x_n$ using a finite number of *symbols* of addition or multiplication; those expressions being, in the ordinary sense of the word, *homogeneous* in $dx_1, dx_2, \ldots, dx_n$. As they are purely symbolical, we will constrain ourselves, each time one meets an addition or multiplication symbol, not to change the order of terms or of factors united by this symbol.

He also defined the notion of *derivative* of a form of degree one (see [28, Chap. 12]), which nowadays is referred to as *differential*[3]:

Being given an expression of Pfaff with n variables

$$\omega = A_1\, dx_1 + A_2\, dx_2 + \ldots + A_n\, dx_n,$$

one calls a *derived expression* the differential expression of the second degree defined by the equality

$$\omega' = dA_1\, dx_1 + dA_2\, dx_2 + \ldots + dA_n\, dx_n.$$

The fundamental property of this derivative is the following one:

Theorem *If a change of variable transforms the expression of Pfaff ω into an expression ϖ, this same change of variables transforms the derived expression ω' into the derived expression ϖ'.*

This theorem allowed the operation of differentiation to be regarded as an *intrinsic* one, applicable to forms defined on *abstract* manifolds, independently of the coordinate systems. Note that in [28], Cartan defined the operation of differentiation only for the forms of degree one. A little later, he extended it in his 1901 paper [29] to forms of arbitrary degree, and he proved the following key theorem:

If ω denotes an arbitrary expression of Pfaff, ω' its derived expression (its bilinear covariant), the expression of the third degree derived from ω' vanishes identically.

With hindsight, one can see here the starting point of *de Rham cohomology*: if V is a differentiable manifold, then its k-th *de Rham cohomology group*, denoted $H^k(V, \mathbb{R})$, is equal to the space of differential forms of degree k with zero differential (the *closed* forms), modulo those which are the derivatives of forms of degree $k - 1$ (the *exact* forms). Let us examine now the reasons why those groups are called *cohomology groups* and why the name of de Rham is attached to them.

---

[3]Frobenius had earlier introduced in his 1877 paper [81] an equivalent notion, under the name of *bilinear covariant* (see [113, page 325]). He was also studying Pfaff's problem.

Chapter 41
de Rham and His Cohomology

In Cartan's works discussed in the previous chapter, the operations on differential forms were used only for *local* studies, related to the resolution of Pfaff systems. Only later, in the 1920s, did Cartan get interested in problems concerning the *global* behaviour of differential forms on manifolds. More specifically, he was interested in the topological structure of the underlying manifolds of Lie groups (which are, by definition, at the same time groups and differentiable manifolds).

In particular, in his 1928 paper [30], he formulated the following conjectures concerning the integration of differential forms on Lie groups:

> If an integral of an exact differential[1] of order p, defined in the closed space $\mathcal{E}$, vanishes for any closed domain of integration with p dimensions then, applying the generalized Stokes formula, it comes from a multiple integral of order $p - 1$ (defined and regular in the whole space).

> If one considers h closed manifolds with p dimensions among which there does not exist any homology, then there exist h integrals of exact differentials such that the square table of the values of those integrals extended to the h manifolds has a determinant different from zero.

Expressed differently, the first conjecture says that *any closed form whose periods[2] vanish is exact,* and the second one that, *given a basis of the p-th homology group with real coefficients and a real number attached to each element of that basis, there exists a closed form of degree p whose periods on that basis constitute precisely that collection of numbers.*

A little later, the Swiss mathematician de Rham proved in his 1931 thesis [49] that these conjectures are in fact true for any closed orientable manifold, not only

---

[1]Nowadays one would call it *closed*, that is, with vanishing differential. An *exact* differential is the differential of another form.

[2]Generalizing the denomination used for closed 1-forms on surfaces, the *period* of a closed p-form ω on the p-dimensional cycle C is the integral $\int_C \omega$.

© Springer International Publishing Switzerland 2016
P. Popescu-Pampu, *What is the Genus?*, Lecture Notes in Mathematics 2162,
DOI 10.1007/978-3-319-42312-8_41

for those underlying Lie groups. Here is the way in which he described the context
of his work, in the 1975 paper [50]:

> At that time, working on a thesis in Analysis Situs, I had studied Poincaré's papers, and
> during a study trip to Paris, where I benefited from the precious advice and encouragements
> of Henri Lebesgue, I got familiarized with the exterior differential calculus[3] by attending
> a course given by Elie Cartan and by reading his Lessons on integral invariants. The few
> mathematicians who then dealt with topology did not know the exterior differential calculus.
> Luckily, those circumstances allowed me to give the first proof of those conjectures which
> Elie Cartan, with his well-known modesty and kindness, called the first propositions of
> Poincaré and then the theorems of de Rham.
>   In fact, Poincaré did not formulate those propositions. [...] Poincaré did not introduce,
> even under other names, the notion of exterior differential. [...] He did not mention Stokes'
> general formula. Nevertheless, in his Lessons on the Geometry of Riemann spaces, whose
> first edition appeared in 1928, E. Cartan refers to the fact that the second differential of
> a differential form is always zero, as a theorem of Poincaré. This result is trivial and
> follows immediately from the definition of this exterior differential. The converse, valid in
> Euclidean space, is not trivial, and in his Lessons on integral invariants, appearing in 1922,
> E. Cartan proves the theorem and its converse mentioning neither Poincaré nor anybody
> else. Today it is this converse which is rather commonly called Poincaré's lemma. But those
> propositions are perfectly enunciated and proved in works of *Volterra*[4] from 1889; they also
> contain the Stokes formula in its general form, as well as—under another name indeed—
> the notion of a harmonic form in Euclidean space. It is clear that Cartan did not know those
> works, otherwise he would not have failed to cite them and to render justice to Volterra.

The key formula which allows one to understand those considerations relating
the notion of homology introduced by Poincaré and the differential forms is indeed
the *generalized Stokes formula*. Nowadays it may be presented very simply in the
following way:

Theorem 41.1 *Let V be a compact, oriented manifold of dimension n, whose
boundary ∂V is oriented by the* outside normals.[5] *Consider a differential form ω
of degree $n - 1$ on V. Let $d\omega$ be the differential of ω. Then:*

$$\int_{\partial V} \omega = \int_V d\omega.$$

The evolution of the ideas leading to the previous formulation, well-adapted to
make the transition from Poincaré's homology to de Rham cohomology, inspired the
recent novel [10] of Michèle Audin. This novel bears some analogy with the present
book, because the main characters of both of them are evolving mathematical ideas:
the Stokes formula and the genus, respectively.

---

[3]That is, the calculus of differential forms of all degrees. It was called *exterior* in reference to the
operation of differentiation, which was qualified "exterior" because it sent the space of forms of a
given degree outside itself.

[4]See for instance: *Opere matematiche*. Vol. I, pages 407 and 422.

[5]That is, the boundary ∂V is oriented so that if one takes at any point of it *an ambient tangent
vector pointing outside* followed by a direct tangent frame for the boundary, then one gets a direct
tangent frame for V.

The notation $d\omega$ for the differential of ω has a great advantage: it allows one to speak about the properties of *the operator d*, the notation with apostrophe making it more difficult to concentrate on the properties of ' ! Thus, Cartan's theorem cited lastly in Chap. 40 may be written as:

$$d \circ d = 0.$$

This formula corresponds to the property (39.2) of the boundary operator acting on chains, which may be written analogously as:

$$\partial \circ \partial = 0.$$

Consequently, if one denotes by Ω^k the vector space of smooth forms of degree k, one gets a chain complex:

$$\cdots \to \Omega^{k-1} \to \Omega^k \to \Omega^{k+1} \to \cdots \tag{41.1}$$

and, therefore, a homology theory, as explained in Chap. 38 in the comments about formula (39.1). Note that, in contrast to that formula, in (41.1) the indices increase in the sense indicated by the arrows. For this reason, one speaks rather about *cohomology*. This also refers to the fact that the *de Rham cohomology groups* $H^k(V, \mathbb{R})$ are *dual* to the homology groups $H_k(V, \mathbb{R})$. Indeed, de Rham's theorems may be reformulated in the following way:

Theorem 41.2 *Assume that V is a closed oriented manifold. Then, the canonical bilinear form*

$$H^k(V, \mathbb{R}) \times H_k(V, \mathbb{R}) \longrightarrow \mathbb{R}$$

defined by the integration of a form on a cycle[6] *is non-degenerate.*

This implies that $H^k(V, \mathbb{R})$ is canonically isomorphic to the dual of $H_k(V, \mathbb{R})$. By combining this with Poincaré's Duality Theorem 39.1, one gets a canonical identification:

$$H_{n-k}(V, \mathbb{R}) \simeq H^k(V, \mathbb{R}). \tag{41.2}$$

In particular, on a closed oriented manifold of dimension 4, the intersection form on $H_2(V, \mathbb{R})$ may be transported by this isomorphism into a symmetric bilinear form on $H^2(V, \mathbb{R})$. It is this bilinear form which was studied by Hodge in his proof of Theorem 34.1, sketched in Chap. 34.

For more details on the development of homology and cohomology theories, one may consult the historical books [62] and [108].

---

[6]By the generalized Stokes formula, this bilinear form is well-defined at the level of Poincaré homology and de Rham cohomology.

Chapter 42
Hodge and the Harmonic Forms

Recall that Riemann used closed paths on his surfaces in order to integrate smooth forms of degree 1 along them. In fact, Riemann did not work with arbitrary such forms, but rather with those which may be written as $f'(z)dz$ in terms of a local holomorphic coordinate z, the function $f(z)$ being also holomorphic. The condition of holomorphicity on $f(z)$ translates to the fact that its real and imaginary parts u, v are both:

- *harmonic*, that is, zeros of the *Laplacian* $\Delta := \dfrac{\partial^2}{\partial x^2} + \dfrac{\partial^2}{\partial y^2}$ relative to the real coordinates (x, y), where $z = x + iy$;
- *conjugated*, that is, that they satisfy the *Cauchy–Riemann* system of equations:

$$\begin{cases} \dfrac{\partial u}{\partial x} = \dfrac{\partial v}{\partial y}, \\ \dfrac{\partial u}{\partial y} = -\dfrac{\partial v}{\partial x}. \end{cases}$$

Hodge succeeded in extending these notions to higher dimensions, this time not on an algebraic variety, but on any smooth manifold endowed with a *Riemannian metric*.[1] It is this last structure which allowed him to generalize both the notion of *conjugation* and that of a *Laplacian operator* to forms of any degree. Therefore one also had a corresponding notion of *harmonic form* in all degrees.

---

[1] In modern terms, a *Riemannian metric* on the smooth manifold V is a smooth field of positive definite quadratic forms on the tangent spaces of V at its various points. At that time, the notion of fiber bundle was just emerging (see Chap. 46), and a Riemannian metric was viewed instead in local coordinates $(x^i)_i$ as an expression $\sum_{i,j} g_{ij} dx^i dx^j$, the multiplication of differentials being commutative, the coefficients g_{ij} being smooth as functions of the variables x^i, symmetric, that is, $g_{ij} = g_{ji}$, and the symmetric matrix $(g_{ij})_{i,j}$ being positive definite for all values of the parameters x^i.

© Springer International Publishing Switzerland 2016
P. Popescu-Pampu, *What is the Genus?*, Lecture Notes in Mathematics 2162,
DOI 10.1007/978-3-319-42312-8_42

Hodge presented his theory systematically in his 1941 treatise [103]. His fundamental theorem was:

Theorem 42.1 *Let V be a closed oriented differentiable manifold, endowed with a Riemannian metric. Then, each de Rham cohomology class of V contains a unique harmonic form.*

Here is what de Rham wrote about this theorem in [51]:

Concerning the theorem ensuring the existence, on a closed manifold with n dimensions, of a closed p-form having preassigned periods, Hodge asked himself whether one could not impose conditions on this p-form which, on a closed Riemann surface, get reduced for $p = 1$ to the condition that the 1-form be harmonic. More precisely, that it be locally the differential of a harmonic function. It was indeed well-known that, on such a surface, there exists only one form of degree 1, having preassigned periods, which is locally the differential of a harmonic function; it is the real part of an abelian differential of the first species.

In order to be able to formulate such conditions on a manifold with $n > 2$ dimensions, it was necessary to introduce a more precise notion than that of a differentiable manifold. Guided by Volterra's works on harmonic functionals, Hodge saw that it was simply convenient to endow the manifold with a Riemannian metric. It allows us to define the operator $*$ [this operator is now called *Hodge's involution*] which associates to any p-form ω the $(n-p)$-form $*\omega$, called the adjoint to ω, whose value on an $(n-p)$-vector is equal to the value of ω on the p-vector with the same norm which is orthogonal to it.

On a Riemann surface, the complex analytic structure determines a Riemannian metric only up to a factor, $ds^2 = \lambda \, |dz|^2$, $z = x + iy$ being a local complex coordinate. But the adjoint of a 1-form $\alpha = a \, dx + b \, dy$, which is $*\alpha = a \, dy - b \, dx$, does not depend on this factor λ and one sees that α is locally the differential of a harmonic function if and only if α and $*\alpha$ are closed.

This led Hodge, still inspired by Volterra's works, to formulate the following definition: in a Riemann space, *a p-form ω is called harmonic if ω and $*\omega$ are closed.*

The applications of the theory to smooth complex manifolds, of which we saw an example in Chap. 34, were done by considering an auxiliary Riemannian metric, adapted in a certain sense to the complex structure. The key property of this metric was its *Kähler*[2] nature.

The Kähler manifolds are still fundamental in algebraic and differential geometry. It was Weil who gave them this name, for the following reason, which he explained in [187]:

In his now classical work on hermitian manifolds [37], Chern had noted by the way the interest which is attached, from the viewpoint of differential geometry, to the hermitian metrics "without torsion", and he had cited with respect to this (p. 112) the rather poorly known work of Kähler [110] in which he was the first to introduce this kind of structure. [...] The study of Hodge's book was not long to bring out the importance of those metrics "without torsion", or, as I suggested to say, Kählerian.

---

[2]A *Kähler metric* on a complex manifold is a Riemannian metric g such that the exterior form of degree 2 defined by $\omega(X, Y) := g(X, iY)$ (where X, Y are any two tangent vectors to the same point) is *closed*.

Let us explain one of the consequences of the hypothesis that a given Riemannian metric on a complex manifold X is Kähler. The complex structure allows us to write in a unique way each differential form α of degree m with complex coefficients as a sum:

$$\alpha = \sum_{p+q=m} \alpha^{p,q},$$

where $\alpha^{p,q}$ can be expressed in any local system of coordinates $(z_1, \ldots, z_n)$ as a sum of differential monomials of the type:

$$a(z_1, \ldots, z_n) dz_{i_1} \wedge \cdots \wedge dz_{i_p} \wedge \overline{dz_{j_1}} \wedge \cdots \wedge \overline{dz_{j_q}}$$

(one says that $\alpha^{p,q}$ is *of type* (p, q)).

Well, *assuming that the metric is Kähler*, if one applics this decomposition to a harmonic form α, then *every form $\alpha^{p,q}$ is still harmonic*. This gives a decomposition:

$$H^m(X, \mathbb{C}) = \bigoplus_{p+q=m} H^{p,q}(X, \mathbb{C}),$$

the isomorphism of Hodge's Theorem 42.1 sending every subspace $H^{p,q}(X, \mathbb{C})$ onto the space of harmonic forms of type (p, q). As the complex vector space $H^m(X, \mathbb{C})$ is the complexification of $H^m(X, \mathbb{R})$, it is invariant by conjugation. One also shows that the complex vector subspaces $H^{p,q}(X, \mathbb{C})$ and $H^{q,p}(X, \mathbb{C})$ are conjugate to each other and thus they have the same dimension. Denoting by:

$$h^{p,q} := \dim_{\mathbb{C}} H^{p,q}(X, \mathbb{C})$$

the (p, q)-*Hodge number*, one gets the following equalities:

$$B_m = \sum_{p+q=m} h^{p,q},$$

$$h^{p,q} = h^{q,p}.$$

An immediate consequence of them is:

Theorem 42.2 *For a smooth closed Kähler manifold, the odd-dimensional Betti numbers are even.*

This theorem constitutes a generalization, different from the one stated in Theorem 39.2, of the fact proved by Riemann (see Chap. 14) that the first Betti number of a closed Riemann surface is even. Indeed, such a surface, that is, a compact smooth complex analytic curve, may always be embedded in a complex projective space.[3] Those spaces admit canonical Kähler structures, which determine by restriction such a structure on the initial complex curve. The same argument shows that any smooth complex projective manifold is Kähler.

Let us return to a general closed Kähler manifold. A simple computation in local coordinates shows that the harmonic forms of type $(p, 0)$ *are exactly the holomorphic forms of degree p.* For instance, for the smooth complex curves, the genus p is equal to $h^{1,0}$; for the smooth complex projective surfaces, the geometric genus p_g is equal to $h^{2,0}$ and the irregularity $p_g - p_a$ is equal to $h^{1,0}$.

From this viewpoint, *the Hodge numbers of a Kähler manifold are a generalization of the genus of a Riemann surface.* However, unlike for Riemann surfaces, starting from complex dimension 2 they are defined only for closed complex manifolds which admit a Kähler metric. In fact:

- In each dimension $n \geq 2$, there exist smooth closed complex manifolds which do not admit such a metric. Indeed, by Theorem 42.2, such a metric does not exist if their first Betti number is odd. A deep theorem, a proof of which may be found in [13, Chap. IV.3], shows that *a smooth closed complex surface admits a Kähler metric if and only if its first Betti number is even.*

- In each dimension $n \geq 2$, there exist smooth closed complex manifolds which admit Kähler metrics, but which are not projective (see the comments about K3 surfaces at the end of Chap. 31).

For more details, one may consult [169, Chap. VIII].

In his 1950 paper [104], Hodge presented a survey of his theory. In it he formulated his famous conjecture, still open, aiming to characterize the rational vector subspace of $H_{2k}(X, \mathbb{Q})$ generated by the fundamental classes of the possibly singular algebraic subvarieties of complex dimension k of the projective manifold X. Slightly reformulated later in order to avoid simple counterexamples, the conjecture may be stated as follows:

If X is a complex projective manifold of dimension n, then the isomorphism (41.2) sends the rational vector subspace of $H_{2k}(X, \mathbb{Q})$ generated by the fundamental classes of k-dimensional algebraic subvarieties of X onto the subspace $H^{n-k,n-k}(X, \mathbb{C}) \cap H^{2(n-k)}(X, \mathbb{Q})$ of $H^{2(n-k)}(X, \mathbb{Q})$.

A nice introduction to Hodge theory was written by Swinnerton-Dyer in [173]. The reader who is eager to learn deeply the present aspects of the Hodge theory of Kähler manifolds may study Voisin's treatise [178].

---

[3]This is a consequence of the Riemann–Roch Theorem 16.3, as we explained in Chap. 16.

Chapter 43
Weil's Conjectures

We saw in Chap. 25 that since his student years, Weil was very interested in the analogies between arithmetic and classical algebraic geometry. That is, between the study of the sets of solutions of systems of Diophantine equations and that of algebraic varieties defined over the field of complex numbers. His most famous paper in this direction is probably [184], published in 1949. He formulated there several conjectures relating geometry over complex numbers and over finite fields, which we now reproduce:

This, and other examples which we cannot discuss here, seem to lend some support to the following conjectural statements, which are known to be true for curves, but which I have not so far been able to prove for varieties of higher dimension.

Let V be a variety without singular points, of dimension n, defined over a finite field k with q elements. Let N_ν be the number of rational points on V over the extension k_ν of k of degree ν. Then we have

$$\sum_{1}^{\infty} N_\nu U^{\nu-1} = \frac{d}{dU} \log Z(U),$$

where $Z(U)$ is a rational function in U, satisfying a functional equation

$$Z(\frac{1}{q^n U}) = \pm q^{n\chi/2} U^\chi Z(U),$$

with χ equal to the Euler–Poincaré characteristic of V (intersection-number of the diagonal with itself on the product $V \times V$).

Furthermore, we have:

$$Z(U) = \frac{P_1(U)P_3(U)\cdots P_{2n-1}(U)}{P_0(U)P_2(U)\cdots P_{2n}(U)},$$

© Springer International Publishing Switzerland 2016
P. Popescu-Pampu, *What is the Genus?*, Lecture Notes in Mathematics 2162,
DOI 10.1007/978-3-319-42312-8_43

with $P_0(U) = 1 - U$, $P_2(U) = 1 - q^n U$, and, for $1 \leq h \leq 2n - 1$:

$$P_h(U) = \prod_{i=1}^{B_h}(1 - \alpha_{h_i} U)$$

where the α_{h_i} are algebraic integers of absolute value $q^{h/2}$.

Finally, let us call the degrees B_h of the polynomials $P_h(U)$ the Betti numbers of the variety V; the Euler–Poincaré characteristic χ is then expressed by the usual formula $\chi = \sum_h (-1)^h B_h$. The evidence at hand seems to suggest that, if $\overline{V}$ is a variety without singular points, defined over a field K of algebraic numbers, the Betti numbers of the varieties $V_\mathfrak{p}$, derived from $\overline{V}$ by reduction modulo a prime ideal $\mathfrak{p}$ in K, are equal to the Betti numbers of $\overline{V}$ (considered as a variety over complex numbers) in the sense of combinatorial topology, for all except at most a finite number of prime ideals $\mathfrak{p}$.

An essential notion in the above extract is that of the *Euler–Poincaré characteristic* χ. We will dedicate Chap. 47 to it. We would only like to note here that viewing $\chi(V)$ as the self-intersection number of the diagonal in the algebraic manifold $V \times V$, as does Weil in the previous quotation, has the advantage of making this notion extendable over any field.

Let us explain this viewpoint on the Euler–Poincaré characteristic for the sake of the reader familiar with vector bundles. Assume that V is a complex manifold. Then the self-intersection number of the diagonal in $V \times V$ is equal to its self-intersection in its normal bundle, which is in turn isomorphic to the tangent bundle of V. Therefore, it is equal to the intersection number of the 0-section with any smooth section of the tangent bundle which meets the 0-section at a finite number of points. But such a section is a smooth vector field on V with a finite number of zeros, and the two sections intersect exactly at the zeros of the vector field. The contribution of each zero to the total intersection number is the *index* of the vector field at that point. By the *Poincaré–Hopf theorem*, the sum of those indices is equal to the Euler–Poincaré characteristic of V.

Weil's conjectures posed in particular the question of developing homological and cohomological theories for algebraic varieties defined over arbitrary fields. They constituted an important program of research until the 1970s, guided among others by Grothendieck, and finalized by Deligne. One may start from Houzel's text [105, Chap. XIII] and Hindry's paper [93] in order to explore their prehistory and from Hartshorne's text [90, Appendix C] in order to explore the history of their proofs.

We will not examine Weil's conjectures any longer. I wanted to mention them in order to present one of the essential reasons why, starting from the 1950s, algebraic geometry, its theories and essential theorems, were reformulated *over any field* or, more generally, over any ring. Such an essential theorem of classical algebraic geometry is that of Riemann–Roch, whose prototype for algebraic curves was stated in Chap. 16. We will next discuss the issues that arise when trying to extend it to higher dimensions.

Chapter 44
Serre and the Riemann–Roch Problem

Since the beginnings of the development of the theory of algebraic surfaces, geometers have tried to prove a generalization of the Riemann–Roch theorem for algebraic curves (see Zariski's account in [195, Chap. IV], as well as Gray's article [84]). Max Noether formulated such a generalization in his 1886 paper [144]. But he could only prove an inequality (I use the notations and the explanations from [32, Chaps. 7 and 35]):

Theorem 44.1 *Let us consider an algebraic surface with numerical genus p_n. Let $|C|$ be a complete and special[1] linear system of genus[2] π, of degree[3] n and dimension r. Denote by r_1 the dimension of the residual system $|K - C|$. Then, one has the following inequality:*

$$r_1 \geq p_n - (\pi - n + r).$$

In spite of various efforts, it remained for a long time difficult to interpret the difference between the two sides of this inequality. Moreover, it was not even clear how to conjecture an analogous theorem in higher dimensions. This seemed to concern the dimensions of complete linear systems $|H|$ and $|K - H|$ on a projective manifold, where H denotes a hypersurface inside it, as well as the arithmetic genus of the manifold, but it was difficult to say more.

Finally, Serre was able to formulate a general conjecture in 1953 using the very recent theory of *sheaf cohomology*, as may be seen in the following extract from the letter [166] he sent to Armand Borel:

---

[1]This means that $|C|$ is contained in the canonical system $|K|$.

[2]This is the arithmetic genus of the curve C, which may be computed using the adjunction formula stated in Theorem 37.2.

[3]This is the number of varying intersection points of two general curves of the system, that is, the number of such points which are not base points of the system.

© Springer International Publishing Switzerland 2016
P. Popescu-Pampu, *What is the Genus?*, Lecture Notes in Mathematics 2162,
DOI 10.1007/978-3-319-42312-8_44

I am now ready to speak about the Riemann–Roch theorem. One has a *compact* manifold X, and a divisor D on X. The question is to say something about the dimension $l(D)$ of the space of functions f, meromorphic on X, such that $(f) \succ -D$. The integer $l(D)$ is finite, as we will see, and it depends only on the *class* of D (for the linear equivalence, where a divisor is called ~ 0 if it is the divisor of a meromorphic function).

One knows that a divisor defines a principal fibre space with fibre $\mathbb{C}^*$, which completely characterizes its class. By adding a 0 to each fibre, one gets a fibration with fibre equal to a vector space of dimension 1, E_D, and one sees immediately that the f satisfying $(f) \succ -D$ correspond bijectively to the *sections* of E_D. Let us agree to denote by $H^q(D)$ the cohomology group H^q of X, with coefficients in the sheaf of germs of sections of E_D, and by $h^q(D)$ the dimension of $H^q(D)$. Therefore, one has:

$$l(D) = h^0(D), \text{ which shows the finiteness of } l(D).$$

For the sequel, it is important to describe directly the sheaf of germs of sections of E_D (denote it by F_D), starting from D: it is simply the subsheaf of the sheaf of all meromorphic functions on X formed by the germs of functions which are (locally) $\succ -D$. At a point not situated on D, it is therefore the sheaf of germs of holomorphic functions; in any case, it is a locally free sheaf of dimension 1 (on $\mathcal{O}_X$).

I will denote by $\chi(D)$ the integer $h^0(D) - h^1(D) + \ldots + (-1)^n h^n(D)$, it is *the [Euler–Poincaré] characteristic of D* (or rather of the sheaf F_D). Of course, at this point, I still do not know that it is well-defined, meaning that the $h^q(D)$'s are *finite*. Riemann–Roch as I conceive it seems to be this:

$\chi(D)$ *is defined for every divisor D, it depends only on the cohomology class x of the divisor D, and may be computed from x and the Chern classes $C_2, \ldots, C_{2n}$ of X by a formula*:

$$\langle P(x, C_2, \ldots, C_{2n}), X \rangle = \chi(D),$$

where P is a polynomial of degree $2n$, independent of n.

The previous statement is still not proved in its entire generality. But I have good hope, at least for algebraic varieties.

In the rest of the letter, Serre proved the Riemann–Roch theorem in the form proposed by him in the case of curves and surfaces. Here is his statement of the theorem for projective surfaces, under a restrictive hypothesis which allowed him to share with Borel the main ideas of the proof:

Let X be a smooth projective surface. If $\chi(0)$ is defined and equal to $p_a + 1$, and if all the irreducible components of the divisor D are without multiple points, one has the formula:

$$\chi(D) = [\ldots] = p_a + 1 + D \cdot (D - K)/2.$$

[$\ldots$] In the case where the surface is Kähler,

$$\chi(D) = \langle x \cdot (C_2 + x)/2 + (C_4 + C_2 \cdot C_2)/12, X \rangle.$$

In the next chapter we will briefly speak about the new viewpoints which are present in this letter. But let us explain first the reason why the previous theorem is a strengthening of the inequality stated in Theorem 44.1. We will do this only when the linear system $|C|$ has no base-points. Then, comparing the language of Castelnuovo and Enriques to that of Serre, one has:

- the dimension r of the complete linear system $|C|$ is equal to $h^0(C) - 1$;
- the dimension r_1 of the residual complete linear system $|K - C|$ is equal to $h^0(K - C) - 1$, which is equal to $h^2(C) - 1$, by a theorem of duality which Serre proved approximately at the same time (it was published for manifolds of arbitrary dimension in [164]), and which states that $H^2(C)$ *is canonically isomorphic to the dual of* $H^0(K - C)$;
- the numerical genus p_n is the arithmetic genus p_a;
- the degree n of C is equal to the self-intersection $C \cdot C$ of the curve C on the surface;
- the genus π of C may be computed by the adjunction formula from Theorem 37.2.

By gathering all those ingredients, one sees that the inequality written in Theorem 44.1 is equivalent to:

$$h^0(C) + h^2(C) \geq p_a + 1 + C \cdot (C - K)/2.$$

But this is indeed a consequence of the equality proved by Serre:

$$h^0(C) + h^2(C) - (p_a + 1 + C \cdot (C - K)/2) = h^1(C) \geq 0.$$

Therefore, one sees that *sheaf cohomology allows one to interpret the difference between the two sides of the inequality of Theorem* 44.1: it is the dimension of $H^1(C)$,[4] the first cohomology group of the sheaf of holomorphic sections of the fibration E_C. This was one of the first successes of the newly born theory of sheaf cohomology. One understands better why the Italian geometers had such difficulties in interpreting this difference!

---

[4] A little later, it became usual to denote this group rather by $H^1(\mathcal{O}_X(C))$, if X is the ambient surface. Here $\mathcal{O}_X(C)$ is the now standard notation for the sheaf denoted by F_C in Serre's letter to Borel.

Chapter 45
New Ingredients

The ingredients appearing in Serre's letter to Borel and which were very recent at that time are: *complex fibrations and their Chern classes, sheafs of holomorphic sections and their cohomology groups*. I will briefly explain the meaning of these notions and relate them to the objects discussed up to now.

A function may be represented geometrically by its graph, which lives in the cartesian product of the source and target. But this construction does not apply to situations in which one has "functions" which take values in distinct spaces at distinct points. This is for instance the case for differential forms on a manifold, which may be seen as "functions" defined on the manifold, whose value at a point belongs to some exterior power of the cotangent space at that point.

A solution for constructing an analog of the graph is then to consider a different target-space for every point of the source-space, and to put all those spaces together, like the fibres in the stem of a plant. Then, one obtains what is called a *fibre bundle*, filled with the target-spaces which are called its *fibres*. When the fibres are vector spaces of the same dimension, one speaks of a *vector bundle*. If, moreover, the dimension of the fibres is 1, then one speaks of *line bundles*. The graph of a function in this generalized sense is a subvariety of the fibre bundle which intersects the fiber corresponding to a point s of the source space S at the image of s by the "function". Such a graph is called a *section* of the fibre bundle, because it may be thought of as an analog of the surface obtained by sectioning the stem of a plant transversally to its fibres.

A rational function of one variable, regarded as a function on the Riemann sphere, is uniquely determined, up to scalar multiplication, by the collection of its roots and poles counted with multiplicities. Such a collection may be seen as a formal sum of points of the complex line. This fact generalizes to the formal sums of hypersurfaces in a smooth compact algebraic variety, called *divisors*, provided that one allows functions in the previous generalized sense (that is, sections of fibre bundles). In fact, as explained by Serre, each divisor D canonically determines a line bundle E_D, endowed with a section whose locus of zeros and poles counted with multiplicities is

© Springer International Publishing Switzerland 2016
P. Popescu-Pampu, *What is the Genus?*, Lecture Notes in Mathematics 2162,
DOI 10.1007/978-3-319-42312-8_45

exactly D, the hypersurfaces contained in the polar locus being counted negatively. An arbitrary divisor D then determines a complete linear system, which is formed by the zero loci, also counted with multiplicities, of the regular sections (that is, the sections without poles) of the line bundle E_D.

Finally, one succeeds to overcome a difficulty encountered by algebraic geometers who wanted to work only with linear systems: there are line bundles which do not admit global regular sections. Therefore the associated linear system is empty. If one knows only to interpret things using linear systems, it must be confusing to feel the presence of a non-trivial linear system which seems empty, but becomes "visible" by taking a convenient "multiple" of it (one may think, for instance, about Castelnuovo's and Enriques' considerations from [32], quoted in Chap. 30).

Let us come back to an arbitrary vector bundle. One may define notions of compatibility between the structure of the fibre space and that of its *base* (the source space of its sections). This allows one to speak about *holomorphic bundles* and about *holomorphic* or *meromorphic sections*. As we have just explained, such a section does not necessarily exist globally, but there are always plenty of them locally.

In order to clarify the problems of *passage from the local to the global*, another type of "function" was introduced around 1950. It associates to each open set of the base the space of holomorphic sections of the vector bundle which are defined on that open set. Then one gets a basic example of *sheaf* of vector spaces.

In general, a *sheaf of vector spaces* is defined by considering a vector space of "sections" for every open set of a topological space, such that one may restrict the sections to a smaller open set and such that one may uniquely glue sections defined on each open set of a covering, whenever their restrictions to the intersections of pairs of open sets coincide.

The reader interested in the historical developments which led to the notion of a sheaf may consult Chorlay's works [42] and [43].

By reformulating ideas of Leray, who was also the first to introduce, in a slightly different form, the notion of a sheaf, Henri Cartan[1] had explained around 1950 how it was possible to associate *cohomology groups* to every sheaf of abelian groups. It was clear that one had to call them "*cohomology*" groups, because:

Theorem 45.1 *If one considers the sheaf of locally constant $\mathbb{R}$-valued functions on a smooth manifold, then one recovers the de Rham cohomology groups.*

An elegant explanation of this result was given by Weil in his 1947 letter [186] to Henri Cartan.

In fact, the de Rham cohomology groups also appear in Serre's letter, via the *Chern classes*. These are de Rham cohomology classes of even degrees canonically associated with every complex vector bundle. They were introduced by Chern in [41] as a measure of the difference between a given bundle and a trivial bundle. As they are cohomology classes, one may add and multiply them. Therefore, it makes

---

[1] He was a son of Elie Cartan.

sense to speak about a *polynomial* in certain Chern classes and in other cohomology classes.

Given a line bundle on the manifold V, one has only a single Chern class $C_1 \in H^2(V, \mathbb{Z})$. It is the Poincaré dual of the homology class of the divisor defined by a meromorphic section of the fibre bundle. More generally, one of the interpretations of the Chern classes of a complex vector bundle is as duals of the homology classes of the loci of zeros and poles of the meromorphic sections of the given bundle and of bundles constructed from it by certain algebraic operations.

The search for a generalization of the Riemann–Roch theorem, and at the same time of a proof of Severi's conjecture from Chap. 37, played an important role in testing these new cohomological tools. In this direction, various important papers, by Zariski [199], Kodaira [120, 121], Kodaira and Spencer [123], and finally Hirzebruch [96] and [97], appeared in quick succession.

In particular, Hirzebruch conjectured and proved an explicit Serre-type formula in terms of Chern classes. But before explaining this in Chap. 49, let us examine a bit further the notions of *fibre bundles* and of *Euler–Poincaré characteristic*, since they are central in Serre's approach to the Riemann–Roch problem.

Chapter 46
Whitney and the Cohomology of Fibre Bundles

The Chern classes appearing in Serre's letter to Borel of Chap. 44 are particular *characteristic classes*, that is, certain cohomology classes measuring the way in which the fibres of a fibre bundle *"fit together over the whole manifold"*, as written by Whitney at the end of the next excerpt from the introduction to his 1937 paper [193]:

> *The Problems.* In many fields of work one is led to the consideration of n-dimensional spaces. A given dynamical system has a certain number of "degrees of freedom"; thus a rigid body, with one point fixed, has three. A line in euclidean space is determined by four "parameters". We therefore consider the positions of the rigid body, or the straight lines, as forming a space of three, or four, dimensions. But when we try to determine the points of the space by assigning to each a set of three, or four, numbers, we are doomed to failure. This is possible for a small region of either space, but not for the whole space at once. The best we can do is to cover the space with such regions, define a coordinate system in each, and state how the coordinate systems are related in any two overlapping regions. They will be related in general by means of differentiable, or analytic, transformations, with non-vanishing Jacobian. Any such space we shall call a differentiable, or analytic, manifold.
>
> For a complete study of such spaces, we must know not only properties of the euclidean n-space E^n, which we may apply in each coordinate system separately, but also properties which arise from the manifold being pieced together from a number of such systems. It is these latter properties, essentially topological in character, which form the subject of the present address.
>
> Suppose we wish to study differential geometry in an n-dimensional manifold M^n. At each point p of M^n, the possible differentials (or "tangent" vectors) form an n-dimensional vector space $V(p)$, the so-called tangent space at p. For topological considerations, it is sufficient to consider vectors of unit length, or directions, which form a sphere $S(p)$ of dimension $n - 1$. This set of spheres forms the *tangent sphere-space* of M^n. Most of our work will be on the problem, how do the spheres fit together over the whole manifold? Suppose M^n is imbedded in a higher dimensional manifold M^m (for instance, euclidean E^m). Then we may consider the normal unit vectors at each point, forming an $(m - n - 1)$-sphere, and thus the *normal sphere-space*.

© Springer International Publishing Switzerland 2016
P. Popescu-Pampu, *What is the Genus?*, Lecture Notes in Mathematics 2162,
DOI 10.1007/978-3-319-42312-8_46

Nowadays, these two fundamental examples of fibre bundles which are *tangent spaces* and *normal spaces* are defined rather as vector bundles. But Whitney looked instead at the associated sphere bundles. He cited papers of Hotelling, Seifert, Stiefel and Threlfall as precursors of his study, mainly for circle bundles over surfaces.

Whitney's basic considerations concerning the notion of an *abstract manifold* (his "*n*-dimensional spaces") illustrate the novelty at that time of this way of thinking.[1]

In this paper, Whitney associated to any sphere bundle certain cohomology classes. Those classes were later called "*Stiefel–Whitney classes*". They were the first *characteristic classes* to be defined and served as models for the construction of the *Chern classes* appearing in the Riemann–Roch formula conjectured by Serre. They are among the first important applications of the idea of cohomology, as a distinct object from homology.

What makes cohomology more adapted to this setting than homology is the following. If $M \to N$ is a smooth map between manifolds, it allows us to pull-back any fibre bundle over N to a fibre bundle over M. It is then natural to try to measure how the fibres "*fit together over the whole manifold*" by objects which have the same *contravariant* behaviour. Well, this is the main difference between *cohomology* classes and *homology* classes: only the first are contravariant while the second are, instead, *covariant*.

---

[1] The modern definition, through atlases, of an abstract smooth manifold of arbitrary dimension, seems to have been published for the first time in Veblen and Whitehead's 1931 paper [176]. It was slightly modified and further explored in Whitney's 1936 paper [192]. One may find details about those papers in Chorlay's commented selection [44] of foundational articles on differential geometry and topology.

Chapter 47
Genus Versus Euler–Poincaré Characteristic

In his 1758 paper [72], Euler published the following theorem:

Theorem 47.1 *If S denotes the number of vertices, A the number of edges and H the number of faces of a convex polyhedron, then $S - A + H = 2$.*

In fact, as shown by the extract of his article reproduced in Fig. 47.1 below, Euler wrote this relation in the form $S + H = A + 2$.

The two equations are, of course, equivalent from a logical viewpoint, but they are not equivalent psychologically. Indeed, if one considers the second way of writing the equation, then nothing indicates that the alternating sum $S - A + H$ would have a special importance. It is a generalization of this expression which was to receive the name of *Euler–Poincaré characteristic*, applied to much more general objects than convex polyhedra in the usual 3-dimensional space.

A relation equivalent to that of Theorem 47.1 had already been obtained by Descartes, but it appeared only in a manuscript, which got lost and was rediscovered in 1860 thanks to a copy made by Leibniz (see the historical comments which accompany [57]). Here is the way in which Descartes formulated this relation:

> If four right plane angles are multiplied by the number of solid angles and if one removes from the product 8 right plane angles, there remains the aggregate of all the plane angles which exist at the surface of the solid body under consideration.

To prove that this statement is equivalent to Euler's theorem is a pleasant exercise, based on the fact that the sum of angles of a polygon with n edges is equal to $2(n - 2)$ "right plane angles". But Descartes' formulation drew even less attention than Euler's to the importance of considering the alternating sum $S - A + H$.

During the nineteenth century, Euler's theorem was generalized to polyhedra which are not necessarily convex, and which possibly have holes. Those investigations showed that one always gets a formula of the type $S - A + H = constant$, this constant being expressible in terms of the number of holes of the polyhedron. This research was full of twists and turns. It inspired Lakatos' famous epistemological

© Springer International Publishing Switzerland 2016
P. Popescu-Pampu, *What is the Genus?*, Lecture Notes in Mathematics 2162,
DOI 10.1007/978-3-319-42312-8_47

PROPOSITIO IV.

§. 33. *In omni folido hedris planis inclufo aggregatum ex numero angulorum folidorum et ex numero hedrarum. binario excedit numerum acierum.*

DEMONSTRATIO.

Scilicet fi pouatur vt hactenus :

numerus angulorum folidorum $=$ S

numerus acierum - - - - $=$ A

numerus hedrarum - - - $=$ H

demonftrandum eft, effe S $+$ H $=$ A $+$ 2.

Fig. 47.1 Euler's formula on polyhedra

book [126]. The reader interested in the details of this history may consult Pont's book [152].

Before Poincaré's article [148], discussed in Chap. 38, work on generalizations of this formula seemed to concern only polyhedra. On the other hand, Poincaré explained in [148, Chaps. 16–17] that in fact it concerns even manifolds:

> This theorem [of Euler] was generalized by the admiral Mr. de Jonquières, to the case of non-convex polyhedra. If a polyhedron forms a closed manifold of two dimensions, whose Betti number is P_1, one will have

$$S - A + F = 3 - P_1.$$

[With the modern definition of Betti numbers, described in Chap. 39, one gets rather $2 - B_1$. In Poincaré's formula, F is the number of faces, denoted H by Euler.]

> The fact that the faces are planar has of course no importance, the theorem applies as well to curvilinear polyhedra; it applies also to the subdivision of an arbitrary closed surface into simply connected regions [for Poincaré, this means that they are homeomorphic to discs]; [...].

> I intend to generalize those results to an arbitrary space.

> Let V be a closed manifold with p dimensions. Subdivide it into a certain number of manifolds v_p with p dimensions; those manifolds will not be closed and their boundaries will be formed by a certain number of manifolds v_{p-1} with $p - 1$ dimensions; [etc.]

> The manifold V may have arbitrary Betti numbers, but I assume specifically that the manifolds $v_p, v_{p-1}, \ldots, v_1$ are simply connected.

> I will call $\alpha_p, \alpha_{p-1}, \ldots, \alpha_1$ and α_0 the numbers of v_p's, of v_{p-1}'s, ..., of v_1's and of v_0's. [...]

> I intend to compute the number

$$N = \alpha_p - \alpha_{p-1} + \alpha_{p-2} - \ldots \mp \alpha_1 \pm \alpha_0.$$

[...] In general, one will have

$$N = P_{p-1} - P_{p-2} + \ldots + P_2 - P_1$$

if p is odd, and

$$N = 3 - P_1 + P_2 - \cdots + P_{p-1}$$

if p is even.

If we observe now that the Betti numbers equally distant from the extremes are equal [as explained in Chap. 39, this *theorem of duality of Poincaré* was presented by him in [148, Chap. 9]], one will see that one must have

$$N = 0$$

if p is odd [...].

Using the language explained in Chap. 39, the number α_k of polyhedra of dimension k is equal to the rank of the groups of chains $C_k(V, \mathbb{Z})$ carried by the fixed polyhedral decomposition.

Afterwards, Poincaré's sign convention was modified. It became standard to define the *Euler–Poincaré characteristic* $\chi(C_\bullet)$ of a finite chain complex $C_\bullet$ (see (39.1)) formed by groups of finite rank as the alternating sum:

$$\sum_k (-1)^k \operatorname{rk} C_k.$$

Then one has the following purely algebraic property:

$$\chi(C_\bullet) = \sum_k (-1)^k \operatorname{rk} H_k. \tag{47.1}$$

In particular, this shows that *the Euler–Poincaré characteristic equals the alternating sum of Betti numbers for all the topological spaces which only involve a finite number of non-zero homology groups of finite rank*, not only for the closed manifolds.

With these sign conventions, one shows that the Euler–Poincaré characteristic is a generalization of the notion of the cardinality of a finite set, which we can view as a closed manifold of dimension 0:

Theorem 47.2

(1) *Assume that the compact topological space X is the union $Y \cup Z$ of two compact subspaces Y, Z, such that there exists a finite polyhedral decomposition of X for which Y and Z are subpolyhedra. Then $\chi(X) = \chi(Y) + \chi(Z) - \chi(Y \cap Z)$.*
(2) *If Y and Z admit finite polyhedral decompositions, then $\chi(Y \times Z) = \chi(Y) \cdot \chi(Z)$.*

In principle, this theorem allows one to compute the Euler–Poincaré characteristic of a space by decomposing it into simpler spaces, which are in turn written as products of yet simpler spaces.

For instance, in this way one may prove the following fundamental relation between the genus and the Euler–Poincaré characteristic of a surface, using Theorem 19.1:

Theorem 47.3 *Let T be a closed, connected and orientable surface of genus $p \geq 0$. Then, $\chi(T) = 2 - 2p$.*

Theorem 47.2 also holds for more general spaces than the topological spaces which admit a polyhedral decomposition. One way to extend this result was initiated by Čech [40]. It amounts to working with the parts of a covering of a space X by open subsets and with all their intersections, rather than with the faces of a polyhedral decomposition. It is this approach which allowed Leray, then Cartan and Serre to define a notion of *cohomology* for sheaves.

More precisely, the k-th cochain group $C^k(\mathcal{U}, \mathcal{F})$ of a sheaf of vector spaces associated with a covering $\mathcal{U}$ by open sets U_i consists of a section $s_{i_0,\ldots,i_k}$ of the sheaf on $U_{i_0} \cap \ldots \cap U_{i_k}$ for each k-tuple of indices $(i_0, \ldots, i_k)$, all these sections being moreover chosen antisymmetrically in the indices. The (co)boundary $\delta : C^k \to C^{k+1}$ is then defined by associating with each intersection $U_{i_0} \cap \ldots \cap U_{i_{k+1}}$ the alternating sum:

$$\sum_{0 \leq j \leq k} (-1)^j s_{i_0,\ldots,\hat{i_j},\ldots,i_{k+1}},$$

the notation "$\hat{i_j}$" meaning that one excludes the index i_j.

One sees in particular that a 0-cochain (s_i) is closed (that is, that $\delta(s_i) = 0$), if and only if the partial sections s_i may be glued into a global section of the sheaf $\mathcal{F}$. This shows that *the 0th cohomology group $H^0(\mathcal{U}, \mathcal{F})$ is isomorphic to the group of global sections of the sheaf $\mathcal{F}$*. The other cohomology groups $H^k(\mathcal{U}, \mathcal{F})$ are therefore generalizations of this group of global sections. As they were defined, they depend on the choice of the covering, but taking the limit for finer and finer coverings allows us to obtain a notion of the cohomology group $H^k(X, \mathcal{F})$ of a sheaf $\mathcal{F}$ defined on a topological space X, which is independent of the choice of open covering.

The reader interested in the way the previous definitions and constructions were developed may consult the correspondence between Henri Cartan and André Weil published by Michèle Audin in [9], especially pages 139–250.

Let us come back to the Riemann–Roch problem discussed in Chap. 44. We saw there that, classically, this problem of extending the Riemann–Roch theorem to higher dimensions concerned the computation of the dimensions of complete linear systems $|D|$. This amounts to calculating dimensions of spaces of global sections of sheaves F_D (we use here the notations of Serre's letter to Borel). In fact, taken separately, those dimensions are more difficult to compute than the Euler–Poincaré characteristic of the sheaf. Indeed, this last invariant satisfies a theorem which recalls the cardinal-type formulae of Theorem 47.2:

Theorem 47.4 *If $\mathcal{G}$ is a subsheaf of $\mathcal{F}$, then:*

$$\chi(\mathcal{F}) = \chi(\mathcal{G}) + \chi(\mathcal{F}/\mathcal{G}).$$

The approach proposed by Serre for proving the Riemann–Roch theorem was based on the iterated use of the previous formula. He could make this strategy work for curves and surfaces. But he lacked several ingredients to reach arbitrary dimensions. It was Hirzebruch who achieved a definition and proof for arbitrary complex manifolds. Then Grothendieck could prove the theorem for any manifold defined on an arbitrary algebraically closed field. Chapters 49 and 50 are dedicated to their approaches.

Before that, let us practice a little with the Euler-Poincaré characteristic, by discovering still another aspect of the genus of a Riemann surface.

Chapter 48
Harnack and Real Algebraic Curves

In his 1876 paper [89], Harnack examined the following question: *given a real alge-braic curve of degree d in the projective plane, how many connected components can it have?* He proved:

Theorem 48.1 *If a real plane algebraic curve of degree d is irreducible as a complex curve, then it has at most $1 + \frac{(d-1)(d-2)}{2}$ connected components, and this number can be attained.*

For instance, a real quartic curve can have at most four connected components. A way to obtain a real quartic curve with this maximal number of connected components is to slightly perturb the union of two ellipses intersecting transversally, as in Fig. 48.1.

More precisely, if $Q_1, Q_2 \in \mathbb{R}[x, y]$ are polynomials of degree 2 defining the two ellipses in the real affine plane with coordinates (x, y), then one may take a quartic defined by an equation of the form $Q_1 \cdot Q_2 + \epsilon = 0$, the sign of ϵ being conveniently chosen.

Harnack's proof of the upper bound was based on Bézout's theorem, and was similar to Cayley's use of it, discussed in Chap. 21.

We will provide another proof for the upper bound, which uses the Riemann surface associated with the curve and basic computations with Euler characteristics.

Denote by $C_{\mathbb{R}}$ the given real curve, contained in the real projective plane $\mathbb{P}^2_{\mathbb{R}}$ with homogeneous coordinates $[x : y : z]$. It is the set of zeros of a homogeneous polynomial $P(x, y, z)$ of degree d. The associated complex curve $C_{\mathbb{C}}$ is the set of zeros of this same polynomial in the *complex* projective plane $\mathbb{P}^2_{\mathbb{C}}$. The irreducibility of the curve $C_{\mathbb{C}}$ translates into the irreducibility of the polynomial P in the ring $\mathbb{C}[X, Y, Z]$.

Consider the associated Riemann surface $T_{\mathbb{C}}$ of $C_{\mathbb{C}}$. One has a holomorphic map $\pi : T_{\mathbb{C}} \to C_{\mathbb{C}}$, which is an isomorphism if and only if $C_{\mathbb{C}}$ is smooth. Otherwise, it is an isomorphism only above the smooth part of $C_{\mathbb{C}}$. Let us denote by $T_{\mathbb{R}}$ the preimage by π of the real curve $C_{\mathbb{R}}$.

© Springer International Publishing Switzerland 2016
P. Popescu-Pampu, *What is the Genus?*, Lecture Notes in Mathematics 2162,
DOI 10.1007/978-3-319-42312-8_48

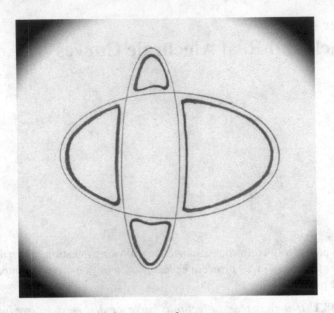

Fig. 48.1 A real quartic curve with four connected components

The conjugation map:

$$[X : Y : Z] \rightarrow [\overline{X} : \overline{Y} : \overline{Z}]$$

acts on $\mathbb{P}^2_{\mathbb{C}}$, and its locus of fixed points is the real projective plane $\mathbb{P}^2_{\mathbb{R}}$. As P is a polynomial with real coefficients, its zero-locus $C_{\mathbb{C}}$ is globally invariant under this action. Therefore, one may restrict the conjugation map to $C_{\mathbb{C}}$, then lift it to $T_{\mathbb{C}}$. The key fact is that $T_{\mathbb{R}}$ *is the fixed point locus of this lift, which is again an involution*:

$$\sigma : T_{\mathbb{C}} \rightarrow T_{\mathbb{C}}.$$

Moreover, this involution is *antiholomorphic*, which means that for any tangent vector X at a point of $T_{\mathbb{C}}$, one has $d\sigma(iX) = -i\,d\sigma(X)$.

This implies that in the neighborhood of any point of $T_{\mathbb{R}}$, there exists a holomorphic local coordinate t on $T_{\mathbb{C}}$ with respect to which σ is given by the usual conjugation $t \rightarrow \bar{t}$. *It is easy to deduce from this that $T_{\mathbb{R}}$ is a disjoint union of finitely many circles.* Harnack's theorem 48.1 is now a consequence of the following intrinsic statement:

Theorem 48.2 *If $T_{\mathbb{C}}$ is a compact, connected Riemann surface of genus $p \geq 1$ and $\sigma : T_{\mathbb{C}} \rightarrow T_{\mathbb{C}}$ is an anti-holomorphic involution of it, with fixed point set $T_{\mathbb{R}}$, then $T_{\mathbb{R}}$ is the disjoint union of at most $p + 1$ circles.*

It is this theorem which we will prove using properties of the Euler-Poincaré characteristic. Denote by S the quotient surface of $T_{\mathbb{C}}$ by the involution σ and let:

$$\rho : T_{\mathbb{C}} \to S$$

be the quotient map. S is a compact and connected smooth surface with boundary. The restriction of ρ to the real locus $T_{\mathbb{R}}$ realizes a diffeomorphism from it to the boundary ∂S. Therefore they have the same number of connected components. Denote this number by b.

Take a triangulation of S and lift it to $T_{\mathbb{C}}$ by the map ρ. Computing Euler–Poincaré characteristics with the aid of those two triangulations, and using the fact that the Euler–Poincaré characteristic of a circle is 0, applied to each boundary component of S, one gets:

$$\chi(T_{\mathbb{C}}) = 2\chi(S). \tag{48.1}$$

Let us glue discs to all such boundary components by homeomorphisms between the boundaries. One obtains a compact connected surface $\bar{S}$ without boundary. The additivity of Theorem 47.2(1) applied to the decomposition $\bar{S} = S \cup D$, where D is the union of the b discs used to close up S, gives:

$$\chi(\bar{S}) = \chi(S) + b. \tag{48.2}$$

Combining Eqs. (48.1) and (48.2) with the fact that $\chi(T_{\mathbb{C}}) = 2(1 - p)$ (see Theorem 47.3), one gets:

$$2(1 - p) = 2(\chi(\bar{S}) - b) \iff b = p + (\chi(\bar{S}) - 1).$$

Now one may use the classification of closed connected surfaces,[1] which shows that their Euler-Poincaré characteristics are at most 2, with equality only for spheres. Therefore $\chi(\bar{S}) - 1 \leq 1$, and the theorem is proved.

One sees, moreover, that the equality $b = p + 1$ holds if and only if $\bar{S}$ is a sphere. In this case, S is homeomorphic to the complement of the interiors of p pairwise disjoint discs lying inside a bigger disc, and one gets the surface $T_{\mathbb{C}}$ simply by identifying the corresponding points of the boundaries of two copies of this complement. In this model, the action of conjugation is simply the exchange of the corresponding points in the two copies of S.

Figure 48.2 illustrates the surface S for the smooth quartic curves of $\mathbb{P}^2_{\mathbb{R}}$ with maximal number of connected components. One may also interpret this drawing as the surface $T_{\mathbb{C}}$ embedded in the usual 3-dimensional space. The three visible holes show that it has genus 3, as expected.

---

[1] In Theorem 19.1 we stated that the orientable surfaces are connected sums of tori. Similarly, it may be shown that the non-orientable surfaces are connected sums of real projective planes.

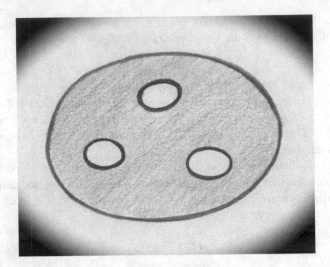

Fig. 48.2   The Riemann surface of the quartic of Fig. 48.1

One may notice that the previous proof is similar to the one we sketched for the Riemann–Hurwitz formula, stated in Theorem 14.1. In both cases, one studies a map between smooth surfaces using the Euler–Poincaré characteristic. In his 1988 paper [177], Viro explained an analogous method to study maps between fairly general topological spaces. The basic insight on which his method is based is that the Euler–Poincaré characteristic is not only a generalization of the cardinality of a finite set, but also an analog of a measure.

The reader wishing to learn more about the topology of *real* algebraic curves in the projective plane may begin by reading Itenberg and Viro's paper [106].

Chapter 49
The Riemann–Roch–Hirzebruch Theorem

Let us come back to the Riemann–Roch problem in higher dimensions.

Part of the difficulty of the conjecture formulated by Serre in his letter to Borel (see Chap. 44) was to find a candidate for the polynomial $P(x, C_2, \ldots, C_{2n})$. In fact, at least in what concerns the arithmetic genus of varieties of dimension at most 6, a candidate had been found by Todd in his 1938 paper [174]. It was stated in a different language, but which was equivalent to the language of Chern classes (see Atiyah's explanations in [8]).

Todd's formulae allowed Hirzebruch to guess the rule of formation of those polynomials, published initially in his 1953 paper [96]. A little later, he could prove the theorem. He sketched his proof in the 1954 paper [97] and he explained it completely in his book [99]. He used very recent results of Thom about the structure of the cobordism ring of smooth closed oriented manifolds. I refer to Hirzebruch's reminiscences [98], as well as to Dieudonné's account [62, Chap. VII.2.A] for details.

Let me describe just the rule of formation of the polynomials, as it led Hirzebruch to a generalization of the notion of genus (see [100]). I follow the notations and the explanations of [99, Chap. I.1].

Assume that B is a commutative ring endowed with a unit element. For every $i \in \mathbb{N}^*$, let us consider a variable p_i of weight i. Set also $p_0 = 1$. Given a sequence $(K_j(p_1, \ldots, p_j))_{j \geq 0}$ of homogeneous polynomials of degrees equal to their indices, with $K_0 = 1$, one denotes briefly:

$$K\left(\sum_{i=0}^{\infty} p_i z^i\right) = \sum_{j=0}^{\infty} K_j(p_1, \ldots, p_j) z^j.$$

© Springer International Publishing Switzerland 2016
P. Popescu-Pampu, *What is the Genus?*, Lecture Notes in Mathematics 2162,
DOI 10.1007/978-3-319-42312-8_49

The sequence is called *multiplicative* if, for two independent sequences $(p_i)_i$ and $(q_i)_i$ of indeterminates:

$$K((\sum_{i=0}^{\infty} p_i z^i) \cdot (\sum_{i=0}^{\infty} q_i z^i)) = K(\sum_{i=0}^{\infty} p_i z^i) \cdot K(\sum_{i=0}^{\infty} q_i z^i).$$

The formal series:

$$K(1 + z) = \sum_{i=0}^{\infty} b_i z^i \in B[[z]]$$

is called the *characteristic series* of the multiplicative sequence. The previous construction establishes a bijection between the set of multiplicative sequences of polynomials with coefficients in the ring B and the multiplicative group of series of $B[[z]]$ whose constant term is 1.

The *Riemann–Roch–Hirzebruch theorem* for line bundles may be stated as follows:

Theorem 49.1 *Let D be a divisor on the complex projective manifold X of dimension n. Let x denote the Chern class of the line bundle E_D. If $(T_k)_{k \in \mathbb{N}^*}$ is the multiplicative sequence of polynomials with rational coefficients and characteristic series $\dfrac{z}{1 - e^{-z}}$, then:*

$$\chi_X(D) = \langle e^x \cdot T_n(C_2, \ldots, C_{2n}), X \rangle.$$

In fact, Hirzebruch also proved an analogous theorem for holomorphic vector bundles of arbitrary rank.

In order to honor the pioneering work [174], Hirzebruch called the multiplicative sequence whose characteristic series is $\dfrac{z}{1 - e^{-z}}$ the *Todd sequence*. Likewise, he called the expression of the arithmetic genus $\chi_X(0)$ given by the second member of the previous equality applied to $x = 0$ the *Todd genus* of X.

Note that Hirzebruch defined the arithmetic genus as the Euler–Poincaré characteristic $\chi_X(0)$, where 0 denotes the zero divisor. Recall that this is the characteristic of the sheaf $\mathcal{O}_X$ of holomorphic functions, regarded as sections of the trivial line bundle. If $p_a(X)$ denotes the definition proposed by Severi (see formula (37.1)), then one has:

$$\chi_X(0) = 1 + (-1)^n p_a(X),$$

which results from the proof of Severi's conjectures by Kodaira (see Chap. 37) and from the general duality theorem of Serre [164].

The advantage of this new definition of the arithmetic genus, which is nevertheless equivalent to the ancient one of Severi, is that one may formulate simpler formulae using it than using Severi's definition. For instance:

Theorem 49.2 *The arithmetic genus à la Hirzebruch is both:*

- additive: *the genus of a disjoint union is the sum of the genera of the members;*
- multiplicative: *the genus of a cartesian product is the product of the genera of the factors.*

Then, Hirzebruch defined a general notion of *genus* for a *class* of manifolds which is left stable by taking disjoint unions and cartesian products. It is an invariant which satisfies the previous properties. As he explained in his book [99], this notion is intimately related to that of multiplicative sequence.

In order to explore more recent developments of this generalization of the notion of genus, one may start from Hirzebruch and Kreck's paper [100].

Chapter 50
The Riemann–Roch–Grothendieck Theorem

Hirzebruch's proof of Theorem 49.1 used in a fundamental way the usual topology of a complex variety, as well as the modern panoply of homology and cohomology groups, vector bundles and their characteristic classes. Therefore, it was not possible to derive from it a proof of an analogous theorem concerning algebraic varieties defined over other fields.

Even to conjecture an analogous statement was difficult, because the theory of Chern classes existed only for complex vector bundles. The challenge was, at the very least, to introduce an analogous notion for varieties defined over finite fields. If they again have to be cohomology classes, one would be forced to develop a cohomology theory for such varieties which would be the analog of de Rham cohomology for complex varieties. One sees therefore that these questions are intimately related to Weil's conjectures presented in Chap. 43.

In his 1955 article [165], Serre established the bases of a cohomology theory of sheaves for varieties defined over an arbitrary field. Here is the beginning of the article:

> One knows that the cohomological methods, and particularly the theory of sheaves, play an increasing role, not only in the theory of functions of several complex variables (see [31]), but also in classical algebraic geometry (let me simply cite the recent works of Kodaira–Spencer on the theorem of Riemann–Roch). The algebraic character of those methods allowed us to hope that it was possible to apply them also to abstract algebraic geometry; the aim of the present memoir is to show that this is indeed the case.

The basic structure allowing one to define a notion of sheaf on those varieties is a natural topology on them, named the *Zariski topology*. In this topology, the closed sets are exactly the algebraic subvarieties.

In what concerns the notion of genus, his theory allowed Serre, among other things, to prove Severi's conjecture presented in Chap. 37, for a smooth projective variety defined over a field of arbitrary characteristic. But he did not obtain a general Riemann–Roch theorem for such varieties.

© Springer International Publishing Switzerland 2016
P. Popescu-Pampu, *What is the Genus?*, Lecture Notes in Mathematics 2162,
DOI 10.1007/978-3-319-42312-8_50

It was Grothendieck who formulated and proved such a theorem, around 1957. He gave a purely algebraic proof of a generalization of the theorem of Riemann–Roch–Hirzebruch, valid *over an algebraically closed field of arbitrary characteristic*. The generalization consisted in the fact that he did not consider only *one* variety, but instead *a family of varieties*. The first proof of his theorem was written in [21] by Borel and Serre. Bott explained the philosophy of the new formulation of the theorem in his commentary [22]:

The advent of sheaf theory has, amongst many things, brought with it a great development of the classical theorem of Riemann–Roch. This paper is devoted to Grothendieck's version of the theorem. Grothendieck has generalized the theorem to the point where not only is it more generally applicable than F. Hirzebruch's version [99], but it depends on a simpler and more natural proof. [...]

[Theorem I :] Let $f : X \rightarrow Y$ be a proper map of quasi-projective varieties, let $\mathcal{F}$ be a coherent sheaf on X, and let the sheaves $R^q f(\mathcal{F})$ on Y be defined by $R^q f(\mathcal{F})_U = H^q(f^{-1}(U); \mathcal{F})$ (U open in X). Then these sheaves are also coherent.

[...] If X is an algebraic variety [...] the group $K(X)$ is defined as follows. Let $F(X)$ denote the free abelian group generated by coherent sheaves over X. Also, if $E : 0 \rightarrow \mathcal{F}_1 \rightarrow \mathcal{F} \rightarrow \mathcal{F}_2 \rightarrow 0$ is a short exact sequence of such sheaves, let $Q(E)$ be the "word" $\mathcal{F} - (\mathcal{F}_1 + \mathcal{F}_2)$ in $F(X)$. Now define $K(X)$ as the quotient of $F(X)$ modulo the subgroup generated by $Q(E)$ as E ranges over the short exact sequences. (We call this construction the K-construction; it can clearly be applied to any category in which short exact sequences are defined.) For example, if p is a point, then $K(p) \approx \mathbb{Z}$ [...], the isomorphism being determined by attaching to a sheaf (which is merely a module over the ground-field in this case) its dimension. This homomorphism is denoted by $ch : K(p) \xrightarrow{\approx} \mathbb{Z}$.

As will be seen, the Riemann–Roch theorem is a comparison statement about $K(X)$ and the Chow ring $A(X)$ which is valid only on non-singular varieties. Accordingly, we will let $\mathfrak{A}$ denote the category of quasi-projective non-singular varieties and their proper maps. On this category $K(X)$ and $A(X)$ partake of both a covariant and a contravariant nature, and it is precisely to complete $K(X)$ to a covariant functor that theorem I is essential.

Grothendieck denotes this covariant homomorphism, induced by a map $f : X \rightarrow Y$ in $\mathfrak{A}$, by $f_!$, and defines it in this way: If $\mathcal{F}$ is a sheaf (coherent, algebraic, will be understood hereafter) then $f_!(\mathcal{F}) \in K(Y)$ shall be the class of the word $\sum_q (-1)^q R^q f(\mathcal{F})$ in $K(Y)$. Because the sum is finite on objects in $\mathfrak{A}$ this operation is well defined, and its linear extension to $F(X)$ is seen to vanish on words of the form $Q(E)$, thus inducing a homomorphism $f_! : K(X) \rightarrow K(Y)$.

The naturality condition $(f \circ g)_! = f_! \circ g_!$ is valid, and follows from the spectral sequence which relates $R^q(f \circ g)$ to $R^s f$ and $R^t g$. Thus the obvious "Euler characteristic" nature of $f_!$ is essential not only for the vanishing of $f_!$ on $Q(E)$, but also for the naturality! Note also that if $f : X \rightarrow p$ is the map onto a point, then $ch\, f_!(\mathcal{F})$ may be identified with $\sum(-1)^q \dim H^q(X; \mathcal{F}) = \chi(X; \mathcal{F})$; and it is an expression of this sort which was evaluated by Hirzebruch in his topological version of the Riemann–Roch theorem by a certain cohomology class. [...]

In the Grothendieck theory, the role of cohomology is taken over by the Chow ring $A(X)$, of cycles under linear equivalence, the product being defined by intersection. On our category, $A(X)$ also has a covariant side to it, namely $f_* : A(X) \rightarrow A(Y)$, defined by the direct image of a cycle. However, f_* is only an additive homomorphism. The contravariant extension of $A(X)$, i.e., $f \rightarrow f^*$ where f^* is induced by the inverse image of a cycle, is of course a ring homomorphism; and these two operations are linked by the permanence law: $f_*(x \cdot f^*(y)) = f_*(x) \cdot y$, $x \in A(X)$, $y \in A(Y)$.

The contravariant properties of $K(X)$ are best brought out with the aid of the following theorem II: Let $K_1(X)$ be the group obtained by applying the K-construction to the category of algebraic vector bundles over $X, X \in \mathfrak{A}$. Also, let $\varepsilon : K_1(X) \to K(X)$ be the homomorphism defined by the operation which assigns to a bundle the sheaf of germs of its sections. Then ε is a bijection.

To a topologist at least, this theorem is reminiscent of the Poincaré duality theorem. In any case, by identifying $K(X)$ with $K_1(X)$ one may induce the obvious (inverse image of a bundle) contravariant extension of $K_1(X)$ to $K(X)$. This homomorphism is denoted by $f_!$. Further, the ring structure of $K_1(X)$ induced by the tensor product of bundles is now also impressed on $K(X)$, and as the authors show, the permanence relation is again valid: $f_!(x \cdot f^!(y)) = f_!(x) \cdot y$ for a map f in $\mathfrak{A}$. This new interpretation of $K(X)$ (i.e., as $K_1(X)$) brings with it also a ring homomorphism $ch : K(X) \to A(X) \otimes \mathbb{Q}$ which is natural on the contravariant side (namely, $ch(f^! x) = f^* ch(x)$) and agrees with our definition of ch on $K(p)$. This function is derived from the Chern character of bundles and can be characterized by: (1) If L is a line bundle over $X \in \mathfrak{A}$, then $ch(L) = e^c = 1 + c + c^2/2! + \cdots$, etc., where $c = c_1(L)$ is the class in $A(X)$ of the zeros of a generic rational section of L; (2) ch is a ring homomorphism; (3) the naturality condition already recorded. [...]

In general, the identification of $K_1(X)$ with $K(X)$ extends the notion of characteristic classes from vector bundles to coherent sheaves. We will, in particular, have need of the Todd-class, which on vector bundles is uniquely characterized by these conditions: (1) If L is a line bundle over an object X in $\mathfrak{A}$, then $T(L) = c/(1 - e^c)$, [...]; (2) T is multiplicative: $T(E + F) = T(E) \cdot T(F)$; (3) $Tf^! = f^*T$ for maps in $\mathfrak{A}$.

This Todd class enters the answer to the following, in our context very natural, question: How does $ch : K(X) \to A(X) \otimes \mathbb{Q}$ behave under the covariant homomorphisms $f_!$ and f_*? The answer to this question is precisely the Riemann–Roch formula of Grothendieck: [...] Let f be a map $X \to Y$, in $\mathfrak{A}$. Then

$$ch\{f_!(x)\} \cdot T(Y) = f_*\{ch(x) \cdot T(X)\},$$

where $x \in K(X)$, and $T(X), T(Y)$ denote the values of the Todd class on the tangent bundles of X and Y respectively.

We see that the theorem of Riemann–Roch–Hirzebruch generalized by Grothendieck may be written almost as briefly as the generalized Stokes formula of Theorem 41.1:

$$ch\{f_!(x)\} \cdot T(Y) = f_*\{ch(x) \cdot T(X)\}.$$

Where is the notion of genus hidden inside it? In the fact that the "good" notion of genus of a variety X is in this context the Euler–Poincaré characteristic of its structure sheaf $\mathcal{O}_X$, which may be generalized as the Euler–Poincaré characteristic of any coherent sheaf, which, in turn, may be generalized as the image of a coherent sheaf by the map $f_!$ induced by a proper morphism f.

To explain conveniently the various notions and constructions which appear in the previous text would require many more pages. I decided therefore to conclude this historical account at this point. The reader may discover other developments of the Riemann–Roch theorem and of the notion of genus (for instance, the *Atiyah–Singer index theorem*) in the appendices of [99].

Concerning Grothendieck, he was in the heart of his period of complete reorganization of algebraic geometry, of its language and problems, an activity partially motivated by the need to create a framework in which to "naturally" prove

Weil's conjectures. One may read his report [88] from 1958 in order to get an idea of his cohomological preoccupations of that time.

If Weil's conjectures have greatly stimulated the mathematicians of the second half of the twentieth century, they were nourished by a deep knowledge of the history. This is what I have tried to show using the numerous quotations of Weil presented throughout this book. Here is a final quote, extracted from the commentary of [183] he included in his collected works:

> [...] I soon became convinced that the constant revisit of the great mathematicians of the past is a source of inspiration no less fertile than the reading of the fashionable authors of the day.

Epilogue

This book is an invitation to consult the writings of the historians of Mathematics, as well as those of mathematicians themselves about the history of their field and, of course, the original works of the discoverers. It is surprising how varied the approaches, the problems and the formulations chosen by different authors are. To look at a phenomenon from various viewpoints or to arrive at an interpretation of the spirit of a period from its writings provide excellent training for the imagination.

But beyond the pleasure which one may take by frequenting the past in this way, is it of any use? I would say that its usefulness is twofold. First, as explained by Feynman [76]:

> Theories of the known, which are described by different physical ideas may be equivalent in all their predictions and are hence scientifically indistinguishable. However, they are not psychologically identical when trying to move from that base into the unknown. For different views suggest different kinds of modifications which might be made and hence are not equivalent in the hypotheses one generates from them in one's attempt to understand what is not yet understood.

Secondly, when one explains mathematics, it is important to be conscious of a plurality of viewpoints, because the mathematical imaginations are also plural. Caring only about the logical structure of a lecture, starting from a definition which comes out of nowhere, continuing with a sequence of lemmas and theorems, is a widespread habit, but which too often carries little meaning.

I believe that knowing the broad lines of the development of a subject, feeling the way one arrives at such a definition, why such a question is important, and the amount of effort spent in order to understand a phenomenon, all of this, not only words, helps to communicate meaning and life. My purpose in writing this story was to try to transmit this belief to the reader.

Let me give several general historical references for the main themes of this story. For a panoramic view of the mathematics of the nineteenth century, one may read Klein's book [118]. For details about the evolution of algebraic geometry, one may consult Dieudonné's book [60] and Houzel's book [105]. For a description of the

© Springer International Publishing Switzerland 2016
P. Popescu-Pampu, *What is the Genus?*, Lecture Notes in Mathematics 2162,
DOI 10.1007/978-3-319-42312-8

evolution of algebraic and geometric topology until the works of Poincaré, one may read Pont's book [152] and for the period between Poincaré and the 1960s, one may consult Dieudonné's book [62]. For historical information on the many topological notions which are essential nowadays, one may consult the collective volume [108].

Some of the papers cited in the book are freely available on one of the websites:

- http://www.math.dartmouth.edu/~euler/
- http://www.mathunion.org/ICM/
- http://www.numdam.org/
- http://www.emani.org/
- http://www.e-periodica.ch/

But, finally, what is the genus?

A very lively idea, which will not let itself be captured by a single definition too soon! I hope to have given the desire to some of my readers to keep exploring this notion and, why not, to contribute to the creation of new interpretations or generalizations of it.

References

1. N.H. Abel, Mémoire sur une propriété générale d'une classe très étendue de fonctions transcendantes, Presented at the French Académie des Sciences in Paris on the 30 October 1826. Republished in *Œuvres complètes de Niels Henrik Abel*, vol. I, ed. by L. Sylow, S. Lie (Grondahl and Son, Christiania, 1881), pp. 145–211
2. N.H. Abel, Lettre à Legendre du 25 Novembre 1828. Republished in *Œuvres complètes de Niels Henrik Abel*, vol. II, ed. by L. Sylow, S. Lie (Grondahl and Son, Christiania, 1881), pp. 271–279
3. N.H. Abel, Démonstration d'une propriété générale d'une certaine classe de fonctions transcendantes. J. Reine Angew. Math. **4**, 200–201 (1829). Republished in *Œuvres complètes de Niels Henrik Abel*, vol. I, ed. by L. Sylow, S. Lie (Grondahl and Son, Christiania, 1881), pp. 515–517
4. S.S. Abhyankar, Historical ramblings in algebraic geometry and related algebra. Am. Math. Mon. **83**(6), 409–448 (1976)
5. Aristotle, *Metaphysics*. Aristotle in 23 volumes, vols. 17, 18. Translation by Hugh Tredennick. Harvard University Press/William Heinemann Ltd, Cambridge/London, 1933/1989. Available at http://www.perseus.tufts.edu/hopper/text?doc=Perseus:text:1999.01.0052
6. M. Artin, Some numerical criteria for contractability of curves on algebraic surfaces. Am. J. Math. **84**, 485–496 (1962)
7. M. Artin, On isolated rational singularities of surfaces. Am. J. Math. **88**, 129–136 (1966)
8. M. Atiyah, Obituary: John Arthur Todd. Bull. Lond. Math. Soc. **30**, 305–316 (1998)
9. M. Audin, *Correspondance Entre Henri Cartan et André Weil (1928–1991)* (Société Mathématique de France, Paris, 2011)
10. M. Audin, *La formule de Stokes, roman* (Cassini, Paris, 2016)
11. R. Ayoub, The lemniscate and Fagnano's contributions to elliptic integrals. Arch. Hist. Exact Sci. **29**(2), 131–149 (1984)
12. H.F. Baker, On some recent advances in the theory of algebraic surfaces. Proc. Lond. Math. Soc. **12**, 1–40 (1913)
13. W.P. Barth, K. Hulek, C.A.M. Peters, A. Van de Ven, *Compact Complex Surfaces*, 2nd Enlarged edn. (Springer, New York, 2004)
14. N. Basbois, L'émergence de la notion de groupe d'homologie. Gaz. Math. **127**, 15–44 (2011)
15. A. Beauville, Surfaces K3, in *Séminaire Bourbaki*, vol. 609 (1982/1983)
16. A. Beauville, et al., *Géométrie des surfaces K3: modules et périodes*. Séminaire Palaiseau 1981. Astérisque, vol. 126 (Société Mathématique de France, Paris, 1985)
17. J. Bernoulli, Constructio Curvae Accesus and Recessus aequabilis, ope rectificationis Curvae cujusdam Algebraicae, addenda nuperae Solutioni mensis Junii. Acta Eruditorum, Sept. 1694.

Republished in *Die Streitschriften von Jacob und Johann Bernoulli*, ed. by H.H. Goldstine (Birkhäuser, Boston, 1991). Jac. Op. LX, pp. 188–192

18. J. Bernoulli, Constructio facilis Curvae Recessus aequabilis a puncto dato, per rectificationem Curvae Algebraicae. Acta Eruditorum, Oct. 1694. Republished in *Opera Omnia Johann Bernoulli*, vol. I, ed. by J.E. Hofmann, G. Olms (Hildesheim, 1968). XIX, pp. 119–122

19. J. Bernoulli, Lectiones Mathematicae de Methodu Integralium, aliisque, conscriptae in usum Ill. Marchionis Hospitalii cum Auctor Parisiis ageret Annis 1691 et 1692, in *Opera Omnia*, vol. III (Lausannae et Genevae, Bousquet, 1742). Reprinted by (Georg Olms Verlag, Hildesheim, 1968), pp. 386–558

20. E. Betti, Sopra gli spazi di un numero qualunque di dimensioni. Ann. Matem. Pura Appl. (2), **4**, 140–158 (1871)

21. A. Borel, J.-P. Serre, Le théorème de Riemann–Roch. Bull. Soc. Math. France **86**, 97–136 (1958)

22. R. Bott, Report on [21]. Math. Rev. MR0116022 (22 #6817)

23. E. Brieskorn, Über die Auflösung gewisser Singularitäten von holomorphen Abbildungen. Math. Ann. **166**, 76–102 (1966)

24. E. Brieskorn, Singularities in the work of Friedrich Hirzebruch, in *Surveys in Differential Geometry*, vol. VII (International Press, Somerville, 2000), pp. 17–60

25. E. Brieskorn, H. Knörrer, *Plane Algebraic Curves* (Birkhäuser Verlag, Boston, 1986). Translation by J. Stillwell of the first German edition of 1981

26. R. Brussee, The canonical class and the C^∞ properties of Kähler surfaces. N. Y. J. Math. **2**, 103–146 (1996)

27. E. Cartan, Le principe de dualité et certaines intégrales multiples de l'espace tangentiel et de l'espace réglé. Bull. Soc. Math. France **24**, 140–177 (1896). Republished in *Œuvres Complètes*. Part II, vol. 1, pp. 265–302

28. E. Cartan, Sur certaines expressions différentielles et sur le problème de Pfaff. Ann. E.N.S. **16**, 239–332 (1899). Republished in *Œuvres Complètes*. Part II, vol. 1, pp. 303–397

29. E. Cartan, Sur l'intégration de certains systèmes de Pfaff de caractère deux. Bull. Soc. Math. France **29**, 233–302 (1901). Republished in *Œuvres Complètes*. Part II, vol. 1, pp. 483–554

30. E. Cartan, Sur les nombres de Betti des espaces de groupes clos. C.R. Acad. Sci. Paris **187**, 196–198 (1928). Republished in *Œuvres Complètes*. Part I, vol. 2 (Gauthier-Villars, Paris, 1952), pp. 999–1001

31. H. Cartan, Variétés analytiques complexes et cohomologie, in *Colloque de Bruxelles* (Georges Thone/Masson and Cie, Liège/Paris, 1953), pp. 41–55

32. G. Castelnuovo, F. Enriques, Sur quelques récents résultats dans la théorie des surfaces algébriques. Math. Ann. **48**, 241–316 (1897)

33. G. Castelnuovo, F. Enriques, Die algebraischen Flächen vom Gesichtspunkt der birationalen Transformationen aus. Enz. Math. Wissensch. III C 6b (B.G. Teubner Verlag, Leipzig, 1915), pp. 674–768

34. G. Castelnuovo, F. Enriques, F. Severi, Max Noether. Math. Ann. **93**, 161–181 (1925)

35. F. Catanese, From Abel's heritage: transcendental objects in algebraic geometry and their algebraization, in [127], pp. 349–394

36. A.L. Cauchy, Mémoire sur les intégrales définies prises entre des limites imaginaires. (De Bure Frères, Paris, 1825). Republished in *Œuvres de Cauchy*, Série II, vol. XV. pp. 41–89

37. A.L. Cauchy, Considérations nouvelles sur les intégrales définies qui s'étendent à tous les points d'une courbe fermée, et sur celles qui sont prises entre des limites imaginaires. C.R. Acad. Sci. Paris **23**, 689–702 (1846).

38. A. Cayley, On the Transformation of Plane Curves. Proc. Lond. Math. Soc. **1**(III), 1–11 (1865/1866)

39. A. Cayley, On the deficiency of certain surfaces. Math. Ann. **3**(4), 526–529 (1871)

40. E. Čech, Théorie générale de l'homologie dans un espace quelconque. Fundam. Math. **19**, 149–183 (1932)

41. S.-S. Chern, Characteristic classes of Hermitian manifolds. Ann. Math. (2) **47**, 85–121 (1946)

42. R. Chorlay, *L'émergence du couple local/global dans les théories géométriques, de Bernhard Riemann à la théorie des faisceaux (1851–1953)*, Thesis, University Paris 7 Diderot, 2007
43. R. Chorlay, From problems to structures: the Cousin problems and the emergence of the sheaf concept. Arch. Hist. Exact Sci. **64**(1), 1–73 (2010)
44. R. Chorlay, *Géométrie et topologie différentielles (1918–1932)* (Hermann, Paris, 2015). A commented selection of articles
45. A. Clebsch, Ueber diejenigen ebenen Curven, deren Coordinaten rationale Functionen eines Parameters sind. J. Reine Angew. Math. **64**, 43–65 (1865)
46. A. Clebsch, Sur les surfaces algébriques. C.R. Acad. Sci. Paris **67**, 1238–1239 (1868)
47. W.K. Clifford, On the canonical form and dissection of a Riemann's surface. Proc. Lond. Math. Soc. **8**(122), 292–304 (1877). Republished in *Mathematical Papers* (Macmillan, London, 1882). Reprinted by Chelsea, New York, 1968
48. D.A. Cox, *Galois Theory* (Wiley Interscience, Hoboken, 2004)
49. G. de Rham, Sur l'analysis situs des variétés à *n* dimensions. J. Math. Pures Appl. (9) **10**, 115–200 (1931)
50. G. de Rham, L'œuvre d'Elie Cartan et la topologie, in *Hommage à Elie Cartan 1869–1951* (Editura Academiei Republicii Socialiste România, Bucarest, 1975), pp. 11–20. Republished in *Œuvres Mathématiques de Georges de Rham* (L'Ens. Math., Univ. de Genève, 1981), pp. 641–650
51. G. de Rham, *Quelques souvenirs des années 1925–1950*. Cahiers du séminaire d'histoire des mathématiques, vol. 1 (Université Pierre et Marie Curie, Paris, 1980), pp. 19–36. Republished in *Œuvres Mathématiques de Georges de Rham* (L'Ens. Math., Univ. de Genève, 1981), pp. 651–668
52. H.P. de Saint-Gervais (pen name of the collective: A. Alvarez, C. Bavard, F. Béguin, N. Bergeron, M. Bourrigan, B. Deroin, S. Dumitrescu, C. Frances, É. Ghys, A. Guilloux, F. Loray, P. Popescu-Pampu, P. Py, B. Sévennec, J.-C. Sikorav), *Uniformization of Riemann surfaces. Revisiting a hundred-year-old theorem*. Eur. Math. Soc. (2016). Translated from the 2011 French edition by R.G. Burns
53. H.P. de Saint-Gervais (pen name of the collective: A. Alvarez, F. Béguin, N. Bergeron, M. Boileau, M. Bourrigan, B. Deroin, S. Dumitrescu, H. Eynard-Bontemps, C. Frances, D. Gaboriau, É. Ghys, G. Ginot, A. Giralt, A. Guilloux, J. Marché, L. Paoluzzi, P. Popescu-Pampu, N. Tholozan, A. Vaugon), *Analysis Situs. Topologie algébrique des variétés.* http://analysis-situs.math.cnrs.fr
54. R. Dedekind, *Theory of Algebraic Integers* (Cambridge University Press, Cambridge, 1996). Commented translation by J. Stillwell of Sur la théorie des nombres entiers algébriques. Bull. Sci. Math. Astron. **11**, 278–288 (1876)
55. M. Demazure, H. Pinkham, B. Teissier (eds.), *Séminaire sur les singulariés des surfaces*. Lecture Notes in Mathematics, vol. 777 (Springer, New York, 1980)
56. R. Descartes, *The Geometry* (Dover Publications, New York, 1954). Translation by D.E. Smith, M.L. Latham of *Géométrie*, Ian Maire marchand libraire à Leyde, 1637, with facsimile in this Dover edition
57. R. Descartes, *Exercices pour les éléments des solides*, in *Collection "Épiméthée"* (Presses Universitaires de France, Paris, 1987). Commented translation by Pierre Costabel of Leibniz's copy of a manuscript of Descartes called *Progymnasmata de Solidorum Elementis*
58. G.C. di Fagnano, Metodo per misurare la Lemniscata, Schediasma II. G. Letterati d'Italia **29**, 258 (1718) and the following ones. Republished in *Produzioni Mathematiche*. II (Pesaro, 1750), pp. 343–348
59. G.C. di Fagnano, Metodo per misurare la Lemniscata, Schediasma II. G. Letterati d'Italia **30**, 87 (1718) and the following ones. Republished in *Produzioni Mathematiche*. II (Pesaro, 1750), pp. 356–368

60. J. Dieudonné, *History of Algebraic Geometry. An Outline of the History and Development of Algebraic Geometry* (Wadsworth Int. Group, Belmont, 1985). Translated by J. D. Sally from the French edition of 1974

61. J. Dieudonné, Emmy Noether and algebraic topology. J. Pure Appl. Algebra **31**, 5–6 (1984)

62. J. Dieudonné, *A History of Algebraic and Differential Topology 1900–1960* (Birkhäuser, Boston/Basel, 1989)

63. S. Donaldson, An application of gauge theory to four-dimensional topology. J. Differ. Geom. **18**(2), 279–315 (1983)

64. S. Donaldson, Irrationality and the h-cobordism conjecture. J. Differ. Geom. **26**(1), 141–168 (1987)

65. P. Du Val, On isolated singularities of surfaces which do not affect the conditions of adjunction. I, II and III. Proc. Camb. Philos. Soc. **30**, 453–459, 460–465, 483–491 (1933/1934)

66. G. Dumas, Sur la résolution des singularités de surfaces. C.R. Acad. Sci. Paris **152**, 682–684 (1911)

67. G. Dumas, Sur les singularités des surfaces. C.R. Acad. Sci. Paris **154**, 1495–1497 (1912)

68. A.H. Durfee, Fifteen characterizations of rational double points and simple critical points. L'Ens. Math. (2) **25**(1–2), 131–163 (1979)

69. F. Enriques, Sur la théorie des équations et des fonctions algébriques d'après l'école géométrique italienne. L'Ens. Math. **23**, 309–322 (1923)

70. F. Enriques, Sur la classification des surfaces algébriques au point de vue des transformations birationnelles. Bull. Soc. Math. France **52**, 602–609 (1924)

71. F. Enriques, *Le superficie algebriche* (Zanichelli, Bologna, 1949)

72. L. Euler, Elementa doctrinae solidorum. Novi Comm. Acad. Sci. Petrop. **4**, 109–140 (1758). Republished in *Opera Omnia*. Serie 1, vol. 26, pp. 71–93

73. L. Euler, Observationes de comparatione arcuum curvarum irrectificabilium. Novi Comm. Acad. Sci. Petrop. **6**, 58–84 (1761). Republished in *Mathematische Werke*. I 20, in *Commentationes Analyticae*. Commentatio 252 (Leipzig and Berlin, 1912), pp. 80–107

74. L. Euler, De formulis integralibus duplicatis. Novi Comm. Acad. Sci. Petrop. **14**, 72–103 (1770). Republished in *Opera Omnia*. Serie 1, vol. 17, pp. 289–315

75. G. Faltings, Endlichkeitssätze für abelsche Varietäten über Zahlkörpern. Invent. Math. **73**(3), 349–366 (1983)

76. R. Feynman, The development of the Space-Time view of quantum electrodynamics. Nobel lecture, 11 December 1965. http://www.nobelprize.org/nobel_prizes/physics/laureates/1965/feynman-lecture.html

77. G. Fischer, *Plane Algebraic Curves*. Student Mathematical Library, vol. 15 (American Mathematical Society, Providence, RI, 2001). Translated from the 1994 German original by L. Kay

78. M.H. Freedman, The topology of four-dimensional manifolds. J. Differ. Geom. **17**(3), 357–453 (1982)

79. R. Friedman, J.W. Morgan, *Smooth Four-Manifolds and Complex Surfaces* (Springer, New York, 1994)

80. R. Friedman, J.W. Morgan, Algebraic Surfaces and Seiberg–Witten Invariants. J. Algebraic Geom. **6**(3), 445–479 (1997)

81. F.G. Frobenius, Über das Pfaffsche Problem. J. Reine Angew. Math. **82**, 230–315 (1877)

82. P. Gario, Resolution of singularities of surfaces by P. Del Pezzo. A mathematical controversy with C. Segre. Arch. Hist. Exact Sci. **40**, 247–274 (1989)

83. P. Gario, Singolarità e Geometria sopra una Superficie nella Corrispondenza di C. Segre a G. Castelnuovo. Arch. Hist. Exact Sci. **43**, 145–188 (1991)

84. J. Gray, The Riemann–Roch theorem and geometry, 1854–1914, in *Proceedings of the International Congress of Mathematicians*, vol. III (Documenta Mathematica, Bielefeld, 1998), pp. 811–822

85. J. Gray, The classification of algebraic surfaces by Castelnuovo and Enriques. Math. Intell. **21**(1), 59–66 (1999)

86. P.A. Griffiths, Variations on a theorem of Abel. Invent. Math. **35**, 321–390 (1976)

87. P.A. Griffiths, The legacy of Abel in algebraic geometry, in [127], pp. 179–205
88. A. Grothendieck, The cohomology theory of abstract algebraic varieties, in *Proceedings of the International Congress of Mathematicians* (American Mathematical Society, Providence, RI, 1958), pp. 103–118
89. A. Harnack, Ueber die Vieltheiligkeit der ebenen algebraischen Curven. Math. Ann. **10**(1), 189–198 (1876)
90. R. Hartshorne, *Algebraic Geometry* (Springer, New York, 1977)
91. F. Hausdorff, *Grundzüge der Mengenlehre* (Veit, Leipzig, 1914)
92. D. Hilbert, Ueber die Theorie der algebraischen Formen. Math. Ann. **36**, 473–534 (1890)
93. M. Hindry, La preuve par André Weil de l'hypothèse de Riemann pour une courbe sur un corps fini, in *Henri Cartan and André Weil mathématiciens du XXᵉ siècle*. Actes des Journées X-UPS 2012 (Éditions de l'École Polytechnique, Palaiseau, 2012), pp. 63–98
94. H. Hironaka, On the arithmetic genera and the effective genera of algebraic curves. Mem. Coll. Sci. Univ. Kyoto. Ser. A. Math. **30**, 177–195 (1957)
95. H. Hironaka, Resolution of singularities of an algebraic variety over a field of characteristic zero: I, II. Ann. Math. **79**(1), 109–326 (1964)
96. F.E.P. Hirzebruch, On Steenrod's reduced powers, the index of inertia, and the Todd genus. Proc. Natl. Acad. Sci. **39**, 951–956 (1953)
97. F.E.P. Hirzebruch, Arithmetic genera and the theorem of Riemann–Roch for algebraic varieties. Proc. Natl. Acad. Sci. **40**, 110–114 (1954)
98. F.E.P. Hirzebruch, The signature theorem: reminiscences and recreation, in *Prospects in Mathematics*. Annals of Mathematics Studies, vol. 70 (Princeton University Press, Princeton, 1971), pp. 3–31
99. F.E.P. Hirzebruch, *Topological Methods in Algebraic Geometry* (Springer, New York, 1978). Translation by R.L.E. Schwarzenberger of the German edition of 1962
100. F.E.P. Hirzebruch, M. Kreck, On the concept of genus in topology and complex analysis. Notices A.M.S. **56**(6), 713–719 (2009)
101. W.V.D. Hodge, The isolated singularities of an algebraic surface. Proc. Lond. Math. Soc. **30**, 133–143 (1930)
102. W.V.D. Hodge, The geometric genus of a surface as a topological invariant. J. Lond. Math. Soc. **8**, 312–319 (1933)
103. W.V.D. Hodge, *The Theory and Applications of Harmonic Integrals* (Cambridge University Press, Cambridge, 1941)
104. W.V.D. Hodge, The topological invariants of algebraic varieties, in *Proceedings of the International Congress of Mathematicians* (American Mathematical Society, Providence, RI, 1950), pp. 182–192
105. C. Houzel, *La géométrie algébrique. Recherches historiques.* (Librairie Scientifique et Technique A. Blanchard, Paris, 2002)
106. I. Itenberg, O. Viro, Patchworking algebraic curves disproves the Ragsdale conjecture. Math. Intell. **18**, 19–28 (1996)
107. C. Jacobi, Considerationes generales de transcendentibus abelianis. J. Reine Angew. Math. **9**, 394–403 (1832)
108. I.M. James (ed.), *History of Topology* (North Holland, Amsterdam, 1999)
109. C. Jordan, Sur la déformation des surfaces. J. Math. Pures Appl. (Journ. de Liouville) (2), **XI**, 105–109 (1866). Republished in *Œuvres de Camille Jordan*, vol. IV (Gauthier-Villars, Paris, 1964), pp. 85–89
110. E. Kähler, Über eine bemerkenswerte Hermitesche Metrik. Abh. Math. Sem. Univ. Hamburg **9**(1), 173–186 (1933)
111. V.J. Katz, The history of differential forms from Clairaut to Poincaré. Hist. Math. **8**, 161–188 (1981)
112. V.J. Katz, Change of variables in multiple integrals: Euler to Cartan. Math. Mag. **55**(1), 3–11 (1982)
113. V.J. Katz, Differential forms – Cartan to de Rham. Arch. Hist. Exact Sci. **33**(4), 321–336 (1985)

114. A. Khovanskii, Newton polyhedra and the genus of complete intersections. Funct. Anal. Appl. **12**(1), 38–46 (1978)

115. S.L. Kleiman, What is Abel's theorem anyway? in [127], pp. 395–440

116. F. Klein, *On Riemann's Theory of Algebraic Functions and their Integrals* (Dover, New York, 1963). Translation by F. Hardcastle of the first German edition published by Teubner, Leipzig, 1882

117. F. Klein, *Lectures on the Icosahedron and the Solution of Equations of the Fifth Degree* (Dover, New York, 2003). Translation by G.G. Morrice of the first German edition published by Teubner, Leipzig, 1884

118. F. Klein, *Development of Mathematics in the 19th Century*. Lie Groups: History, Frontiers and Applications, vol. IX (Mathematical Science Press, Brookline, 1979). Translation of Vol. I of the first German edition published by Springer, 1926

119. H. Kneser, Geschlossen Flächen in dreidimensionalen Mannigfaltigkeiten. Jahresb. Deutschen Math. Ver. **38**, 248–260 (1929)

120. K. Kodaira, On the theorem of Riemann–Roch for adjoint systems on Kählerian varieties. Proc. Natl. Acad. Sci. **38**, 522–533 (1952)

121. K. Kodaira, Arithmetic genera of algebraic varieties. Proc. Natl. Acad. Sci. **38**, 527–533 (1952)

122. K. Kodaira, On the structure of compact complex analytic surfaces I. Am. J. Math. **86**, 751–798 (1964). II, Am. J. Math. **88**, 682–721 (1966). III, Am. J. Math. **90**, 55–83 (1969). IV, Am. J. Math. **90**, 1048–1066 (1969)

123. K. Kodaira, D.C. Spencer, On arithmetic genera of algebraic varieties. Proc. Natl. Acad. Sci. **39**, 641–649 (1953)

124. P. Kronheimer, T. Mrowka, The genus of embedded surfaces in the projective plane. Math. Res. Lett. **1**, 797–808 (1994)

125. O. Labs, *Hypersurfaces with many singularities – history, constructions, algorithms, visualization*. Thesis, University of Mainz, 2005

126. I. Lakatos, *Proofs and Refutations* (Cambridge University Press, Cambridge, 1976)

127. O.A. Laudal, R. Piene (eds.), *The Legacy of Niels Henrik Abel* (Springer, New York, 2004)

128. A.-M. Legendre, *Traité des fonctions elliptiques*, Tome Premier (Huzard-Courcier, Paris, 1825)

129. S.M. Lane, Topology becomes algebraic with Vietoris and Noether. J. Pure Appl. Algebra **39**, 305–307 (1986)

130. M. Merle, B. Teissier, Conditions d'adjonction, d'après Du Val, in [55], pp. 230–245

131. J. Milnor, On manifolds homeomorphic to the 7-sphere. Ann. Math. (2) **64**, 399–405 (1956)

132. J. Milnor, A unique decomposition theorem for 3-manifolds. Am. J. Math. **84**, 1–7 (1962)

133. J. Milnor, Towards the Poincaré conjecture and the classification of 3-manifolds. Not. Am. Math. Soc. **50**(10), 1226–1233 (2003)

134. J. Milnor, Topology through the centuries: low dimensional manifolds. Bull. Am. Math. Soc. (N.S.) **52**(4), 545–584 (2015)

135. A.F. Möbius, Theorie der elementaren Verwandtschaften. Ber. Verh. Sachs. **15**, 18–57 (1863)

136. E.E. Moise, Affine structures in 3-manifolds. V. The triangulation theorem and Hauptvermutung. Ann. Math. (2) **56**, 96–114 (1952)

137. G.H. Moore, The emergence of open sets, closed sets, and limit points in analysis and topology. Hist. Math. **35**(3), 220–241 (2008)

138. D. Mumford, *Curves and Their Jacobians* (The University of Michigan Press, Ann Arbor, 1975). Republished as an Appendix of *The Red Book of Varieties and Schemes*. Lecture Notes in Mathematics, vol. 1358 (Springer, Berlin, 1999)

139. C.G. Neumann, *Vorlesungen über Riemann's Theorie der Abelschen Integralen* (Teubner, Leipzig, 1865)

140. I. Newton, *Analysis per quantitatum series, fluxiones, ac differentias; cum enumeratione linearum tertii ordinis*, ed. by W. Jones, London (1711). English translation: *Enumeration of lines of the third order, generation of curves by shadows, organic description of curves,*

and construction of equations by curves. transl. ed. by C.R.M. Talbot, H.G. Bohn, London, 1860

141. M. Noether, Zur Theorie des eindeutigen Entsprechens algebraischer Gebilde von beliebig vielen Dimensionen. I. Math. Ann. **2**, 293–316 (1870)

142. M. Noether, Zur Theorie des eindeutigen Entsprechens algebraischer Gebilde von beliebig vielen Dimensionen. II. Math. Ann. **8**, 495–533 (1875)

143. M. Noether, Rationale Ausführungen der Operationen in der Theorie der algebraischen Funktionen. Math. Ann. **23**, 311–358 (1883)

144. M. Noether, Extension du théorème de Riemann–Roch aux surfaces algébriques. C.R. Acad. Sci. Paris **103**, 734–737 (1886)

145. E. Noether, Idealtheorie in Ringbereichen. Math. Ann. **83**, 24–66 (1921)

146. G. Perelman, The entropy formula for the Ricci flow and its geometric applications. (2002). ArXiv:math.DG/0211159

147. É. Picard, G. Simart, *Théorie des fonctions algébriques de deux variables indépendantes.* Tome I et II (Gauthier-Villars, Paris, 1897/1906). Reprinted by Chelsea Publishing Company, 1971

148. H. Poincaré, Analysis situs. J. École Polytech. **1**, 1–121 (1895). Republished in *Œuvres de Henri Poincaré*, vol. VI (Gauthier-Villars, Paris, 1953), pp. 193–288

149. H. Poincaré, Complément à l'Analysis Situs. Rend. Circ. Matem. Palermo **13**, 285–343 (1899). Republished in *Œuvres de Henri Poincaré*, vol. VI (Gauthier-Villars, Paris, 1953), pp. 290–337

150. H. Poincaré, Sur les propriétés arithmétiques des courbes algébriques. J. Math. Pures Appl. (Journ. de Liouville) (V), **7**, 161–233 (1901). Republished in *Œuvres de Henri Poincaré*, vol. V (Gauthier-Villars, Paris, 1950), pp. 483–550

151. H. Poincaré, Cinquième Complément à l'Analysis Situs. Rend. Circ. Matem. Palermo **18**, 45–110 (1904). Republished in *Œuvres de Henri Poincaré*, vol. VI (Gauthier-Villars, Paris, 1953), pp. 435–498

152. J.-C. Pont, *La topologie algébrique des origines à Poincaré* (Presses Universitaires de France, Paris, 1974)

153. P. Popescu-Pampu, La bande que "tout le monde connaît". Images des Mathématiques (CNRS, 2010). http://images.math.cnrs.fr/La-bande-que-tout-le-monde-connait.html

154. P. Popescu-Pampu, Idéalisme radical. Images des Mathématiques (CNRS, 2011). http://images.math.cnrs.fr/Idealisme-radical.html

155. P. Popescu-Pampu, Qu'est-ce que le genre?, in *Histoires de Mathématiques.* Actes des Journées X-UPS, vol 2011 (Éditions de l'École Polytechnique, Palaiseau, 2011), 55–198

156. P. Popescu-Pampu, La dualité de Poincaré. Images des Mathématiques (CNRS, 2012). http://images.math.cnrs.fr/La-dualite-de-Poincare.html

157. V. Puiseux, Recherches sur les fonctions algébriques. J. Math. Pures Appl. (Journ. de Liouville) **15**, 365–480 (1850)

158. T. Radó, Über den Begriffe der Riemannsche Fläche. Acta Litt. Sci. Szeged **2**, 101–121 (1925)

159. M. Reid, Chapters on algebraic surfaces, in *Complex Algebraic Geometry*, ed. by J. Kollár (American Mathematical Society, Providence, RI, 1997), pp. 3–159

160. B. Riemann, *Grundlagen für eine allgemeine Theorie der Funktionen einer veränderlicher komplexer Grösse* (Inauguraldissertation, Göttingen, 1851). French translation: *Principes fondamentaux pour une théorie générale des fonctions d'une grandeur variable complexe*, in *Œuvres mathématiques de Riemann.*, transl. L. Laugel (Gauthier-Villars, Paris, 1898), pp. 2–60. Reprinted by J. Gabay, Sceaux, 1990

161. B. Riemann, Theorie der Abelschen Functionen. J. Reine Angew. Math. **54**, 115–155 (1857). French translation: *Théorie des fonctions abéliennes.* Dans *Œuvres mathématiques de Riemann*, transl. L. Laugel (Gauthier-Villars, Paris, 1898), pp. 89–164. Reprinted by J. Gabay, Sceaux, 1990

162. G. Roch, Ueber die Anzahl der willkurlichen Constanten in algebraischen Functionen. J. Reine Angew. Math. **64**, 372–376 (1865)

163. N. Schappacher, Some milestones of lemniscatomy, in *Algebraic Geometry*, ed, by S. Sertöz. Lecture Notes in Pure and Applied Maths. Series, vol. 193 (Marcel Dekker, New York, 1993)

164. J.-P. Serre, Un théorème de dualité. Commun. Math. Helv. **29**(1), 9–26 (1955). Republished in *Œuvres* I (Springer, New York, 2003), pp. 292–309

165. J.-P. Serre, Faisceaux algébriques cohérents, Ann. Math. **61**, 197–278 (1955). Republished in *Œuvres*, vol. I (Springer, New York, 2003), pp. 310–391

166. J.-P. Serre, Lettre à Armand Borel du 16 avril 1953, in *Œuvres* I (Springer, New York, 2003), pp. 243–250

167. F. Severi, Fondamenti per la geometria sulle varietà algebriche. Rend. Circolo Mat. Palermo **28**, 33–87 (1909)

168. F. Severi, Fondamenti per la geometria sulle varietà algebriche. II. Ann. Mat. Pura Appl. (4) **32**, 1–81 (1951)

169. I.R. Shafarevich, *Basic Algebraic Geometry, vol. 2. Schemes and Complex Manifolds* 2nd edn. (Springer, New York, 1994)

170. P. Slodowy, Groups and special singularities, in *Singularity Theory (Trieste, 1991)* (World Scientific Publishing, Singapore, 1995), pp. 731–799

171. A.I. Smadja, *La lemniscate de Fagnano et la multiplication complexe* (2004). halshs-00456361

172. J. Stillwell, *Mathematics and its History*, 2nd edn. (Springer, New York, 2002)

173. P. Swinnerton-Dyer, An outline of Hodge theory, in *Algebraic Geometry. Oslo 1970* (Proceedings of the Fifth Nordic Summer School in Mathematics) (Wolters-Noordhoff, Groningen, 1972), pp. 277–286

174. J.A. Todd, The arithmetical invariants of algebraic loci. Proc. Lond. Math. Soc. **43**, 190–225 (1938)

175. R. Vanden Eynde, Historical evolution of the concept of homotopic paths. Arch. Hist. Exact Sci. **45**(2), 127–188 (1992)

176. O. Veblen, J.H.C. Whitehead, A set of axioms for differential geometry. Proc. Natl. Acad. Sci. **17**(10), 551–561 (1931). With an Erratum on page 660

177. O. Viro, Some integral calculus based on Euler characteristic, in *Topology and geometry – Rohlin Seminar*. Lecture Notes in Mathematics, vol. 1346 (Springer, Berlin, 1988), pp. 127–138

178. C. Voisin, *Hodge Theory and Complex Algebraic Geometry. I., II.* Cambridge Studies in Advanced Mathematics, vol. 76 (Cambridge University Press, Cambridge, 2002). Translation from the French by Leila Schneps

179. A. von Brill, M. Noether, Ueber die algebraischen Functionen und ihre Anwendung in der Geometrie. Math. Ann. **7**(2–3), 269–310 (1874)

180. A. von Brill, M. Noether, Die Entwicklung der Theorie der algebraischen Functionen in älterer und neuer Zeit. Jahresber. Deutsch. Math. Verein. **3**, 107–600 (1894)

181. R.J. Walker, Reduction of the singularities of an algebraic surface. Ann. Math. **36**(2), 336–365 (1935)

182. C.T.C. Wall, *Singular Points of Plane Curves*. London Mathematical Society Student Texts, vol. 63 (Cambridge University Press, Cambridge, 2004)

183. A. Weil, L'arithmétique sur les courbes algébriques. Acta Math. **52**, 281–315 (1928). Republished in *Œuvres scientifiques I* (Springer, New York, 1979), pp. 11–45

184. A. Weil, Numbers of solutions of equations in finite fields. Bull. Am. Math. Soc. (VI) **55**, 497–508 (1949). Republished in *Œuvres scientifiques I* (Springer, New York, 1979), pp. 399–410

185. A. Weil, Number theory and algebraic geometry, in *Proceedings of the International Congress of Mathematicians* (1950), pp. 90–100. Republished in *Œuvres scientifiques I* (Springer, New York, 1979), pp. 442–452

186. A. Weil, Letter to Henri Cartan from 18 January 1947, in *Œuvres scientifiques II* (Springer, New York, 1979), pp. 45–47

187. A. Weil, Comments on the article "Sur la théorie des formes différentielles attachées à une variété analytique complexe", in *Œuvres scientifiques I* (Springer, New York, 1979), pp. 562–564

188. A. Weil, Comments on "On the moduli of Riemann surfaces; Final report on contract AF18(603)-57", in *Œuvres scientifiques II* (Springer, New York, 1979), pp. 545–547

189. A. Weil, Riemann, Betti and the birth of Topology. Arch. Hist. Exact Sci. **20**, 91–96 (1979)

190. A. Weil, Sur les origines de la géométrie algébrique. Comput. Math. **44**, 395–406 (1981)

191. H. Weyl, *The Concept of a Riemann Surface*, 3rd edn. (Addison-Wesley, Reading, MA, 1955). Translation by G.R. Maclane of the first German edition of 1913

192. H. Whitney, Differentiable manifolds. Ann. Math. **37**(3), 645–680 (1936)

193. H. Whitney, Topological properties of differentiable manifolds. Bull. Am. Math. Soc. **43**, 785–805 (1937)

194. A. Wiles, Modular elliptic curves and Fermat's last theorem. Ann. Math. (2) **141**(3), 443–551 (1995)

195. O. Zariski, *Algebraic Surfaces* (Springer, Berlin, Heidelberg, 1935). Reprinted in 1971 with appendices by S.S. Abhyankar, J. Lipman and D. Mumford

196. O. Zariski, Polynomial ideals defined by infinitely near base points. Am. J. Math. **60**(1), 151–204 (1938)

197. O. Zariski, Normal varieties and birational correspondences. Bull. A.M.S. **48**, 402–413 (1942)

198. O. Zariski, Reduction of the singularities of algebraic three dimensional varieties. Ann. Math. (2) **45**, 472–542 (1944)

199. O. Zariski, Complete linear systems on normal varieties and a generalization of a lemma of Enriques-Severi. Ann. Math. (2) **55**, 552–592 (1952)

200. H.G. Zeuthen, Études géométriques de quelques-unes des propriétés de deux surfaces dont les points se correspondent un à un. Math. Ann. **4**, 21–49 (1871)

Index

Abel, 21, 23, 28, 45, 82
  his motivations, 25
Abhyankar, 75
adjoint
  curve, 65
  surface, 83, 99
adjunction, 98
  formula, 115
algebraic
  curve, 11
  function, 10, 28
analysis situs, 39, 40, 117
analytic continuation, 29
antiholomorphic
  involution, 156
Aristotle, 1
Artin, 100
Atiyah, 159
Audin, 130, 152
Ayoub, 18

Bézout
  theorem, 64, 155
Baker, 94
Barth, 95, 106
Basbois, 120
base-point, 91
Beauville, 94
Bernoulli, 9, 11, 73
Betti, 39, 117, 119
birational
  equivalence, 41, 87
  invariance, 85
  transformation, 88

blow up, 56, 66, 88, 89
Borel, 139
Bott, 164
boundary
  of a chain, 121
  orienting it, 130
branch of a curve, 33
Brieskorn, 5, 33, 66, 101
Brill, 75, 82, 91, 93
Brussee, 105
bundle
  fibre, 143, 147
  line, 143
  vector, 143
Burali-Forti, 125

canonical
  series, 91
  system, 91
Cartan
  Elie, 125, 129, 130
  Henri, 144, 152
Castelnuovo, 66, 83, 85, 86, 91–93, 113, 141
Catanese, 49
Cauchy, 27, 31, 35, 69, 119
Cauchy–Riemann
  equations, 29, 133
Cayley, 60, 63, 83–85, 109, 113
chain
  in a manifold, 121
  its boundary, 121
chain complex, 122
characteristic
  as generalized cardinality, 151

© Springer International Publishing Switzerland 2016
P. Popescu-Pampu, *What is the Genus?*, Lecture Notes in Mathematics 2162,
DOI 10.1007/978-3-319-42312-8

Index

Euler–Poincaré, 40, 138, 140, 149–151,
160, 164
function of Hilbert, 110
series, 160
Chern
character, 165
class, 140, 144
Chorlay, 144, 148
Chow ring, 164
class
characteristic, 148
Chern, 140, 144
fundamental, 123
Stiefel–Whitney, 148
classification
for Aristotle, 1
of algebraic curves, 93
of algebraic surfaces, 93, 94
of conics, 7
of cubics, 7
of quadrics, 87
of real closed surfaces, 117
Clebsch, 59, 63, 65, 82, 85, 93, 109
Clifford, 52, 53
cohomology, 131
contrast with homology, 148
de Rham, 127, 131
sheaf, 141, 144, 152
complete
linear series, 49, 82
linear system, 82
conditions of adjunction, 99
conic section, 7
conjecture
Van de Ven, 105
Cartan, 129
Hodge, 136
Mordell, 77
Poincaré, 39
Severi, 115, 145
Thom, 77
Weil, 137, 163
connected sum, 54
connection order, 38
contraction, 88
contravariant
behaviour, 148, 164
covariant
behaviour, 148, 164
Frobenius' bilinear, 127
covering
universal, 69, 120
Cox, 18

Cramer, 63
critical
point, 28
set, 28
curve
adjoint, 65
algebraic, 11
geometric, 6
mechanical, 5, 11
rational, 44
transcendental, 11
cusp, 60
cyclic point, 60

d'Alembert, 74
de Jonquières, 150
de Rham, 129, 134
de Saint-Gervais, 70, 124
Dedekind, 110
deficiency, 63
degree, 6, 7
of a covering, 39
of a divisor, 82
of a linear series, 92
Deligne, 138
Descartes, 5, 74, 149
determination
of an algebraic function, 29
Dieudonné, 120, 159, 167, 168
differential
of a form, 127
dimension
and Hilbert's function, 111
of a linear series, 92
Diophantus, 9, 71, 74
divisor, 143
effective, 49, 82
Donaldson, 105, 106
double point, 60
for Cayley, 63
ordinary, 60
Du Val, 99
singularities, 101
dual graph, 100
duality
Poincaré, 123, 151
Serre, 141
Dumas, 98
Durfee, 101
Dyck, 117

elementary loops, 31
elliptic function, 20

LECTURE NOTES IN MATHEMATICS  Springer

Editors in Chief: J.-M. Morel, B. Teissier;

Editorial Policy

1. Lecture Notes aim to report new developments in all areas of mathematics and their applications – quickly, informally and at a high level. Mathematical texts analysing new developments in modelling and numerical simulation are welcome.

   Manuscripts should be reasonably self-contained and rounded off. Thus they may, and often will, present not only results of the author but also related work by other people. They may be based on specialised lecture courses. Furthermore, the manuscripts should provide sufficient motivation, examples and applications. This clearly distinguishes Lecture Notes from journal articles or technical reports which normally are very concise. Articles intended for a journal but too long to be accepted by most journals, usually do not have this "lecture notes" character. For similar reasons it is unusual for doctoral theses to be accepted for the Lecture Notes series, though habilitation theses may be appropriate.

2. Besides monographs, multi-author manuscripts resulting from SUMMER SCHOOLS or similar INTENSIVE COURSES are welcome, provided their objective was held to present an active mathematical topic to an audience at the beginning or intermediate graduate level (a list of participants should be provided).

   The resulting manuscript should not be just a collection of course notes, but should require advance planning and coordination among the main lecturers. The subject matter should dictate the structure of the book. This structure should be motivated and explained in a scientific introduction, and the notation, references, index and formulation of results should be, if possible, unified by the editors. Each contribution should have an abstract and an introduction referring to the other contributions. In other words, more preparatory work must go into a multi-authored volume than simply assembling a disparate collection of papers, communicated at the event.

3. Manuscripts should be submitted either online at www.editorialmanager.com/lnm to Springer's mathematics editorial in Heidelberg, or electronically to one of the series editors. Authors should be aware that incomplete or insufficiently close-to-final manuscripts almost always result in longer refereeing times and nevertheless unclear referees' recommendations, making further refereeing of a final draft necessary. The strict minimum amount of material that will be considered should include a detailed outline describing the planned contents of each chapter, a bibliography and several sample chapters. Parallel submission of a manuscript to another publisher while under consideration for LNM is not acceptable and can lead to rejection.

4. In general, **monographs** will be sent out to at least 2 external referees for evaluation.

   A final decision to publish can be made only on the basis of the complete manuscript, however a refereeing process leading to a preliminary decision can be based on a pre-final or incomplete manuscript.

   Volume Editors of **multi-author works** are expected to arrange for the refereeing, to the usual scientific standards, of the individual contributions. If the resulting reports can be forwarded to the LNM Editorial Board, this is very helpful. If no reports are forwarded or if other questions remain unclear in respect of homogeneity etc, the series editors may wish to consult external referees for an overall evaluation of the volume.

5. Manuscripts should in general be submitted in English. Final manuscripts should contain at least 100 pages of mathematical text and should always include

   – a table of contents;
   – an informative introduction, with adequate motivation and perhaps some historical remarks: it should be accessible to a reader not intimately familiar with the topic treated;
   – a subject index: as a rule this is genuinely helpful for the reader.
   – For evaluation purposes, manuscripts should be submitted as pdf files.

6. Careful preparation of the manuscripts will help keep production time short besides ensuring satisfactory appearance of the finished book in print and online. After acceptance of the manuscript authors will be asked to prepare the final LaTeX source files (see LaTeX templates online: https://www.springer.com/gb/authors-editors/book-authors-editors/manuscriptpreparation/5636) plus the corresponding pdf- or zipped ps-file. The LaTeX source files are essential for producing the full-text online version of the book, see http://link.springer.com/bookseries/304 for the existing online volumes of LNM). The technical production of a Lecture Notes volume takes approximately 12 weeks. Additional instructions, if necessary, are available on request from lnm@springer.com.

7. Authors receive a total of 30 free copies of their volume and free access to their book on SpringerLink, but no royalties. They are entitled to a discount of 33.3 % on the price of Springer books purchased for their personal use, if ordering directly from Springer.

8. Commitment to publish is made by a *Publishing Agreement*; contributing authors of multiauthor books are requested to sign a *Consent to Publish form*. Springer-Verlag registers the copyright for each volume. Authors are free to reuse material contained in their LNM volumes in later publications: a brief written (or e-mail) request for formal permission is sufficient.

Addresses:
Professor Jean-Michel Morel, CMLA, École Normale Supérieure de Cachan, France
E-mail: moreljeanmichel@gmail.com

Professor Bernard Teissier, Equipe Géométrie et Dynamique,
Institut de Mathématiques de Jussieu – Paris Rive Gauche, Paris, France
E-mail: bernard.teissier@imj-prg.fr

Springer: Ute McCrory, Mathematics, Heidelberg, Germany,
E-mail: lnm@springer.com

Printed in the United States
By Bookmasters